Space Education & Strategic Applications

Also from Westphalia Press
westphaliapress.org

SE:SA SPACE EDUCATION & STRATEGIC APPLICATIONS

VOLUME 6, NUMBER 1 • JANUARY 2026

Dr. Kristen Miller and Dr. Gary Deel, *Co-Editors-in-Chief*

Westphalia Press
An imprint of Policy Studies Organization

Space Education & Strategic Applications
Vol. 6, No. 1 • January 2026

Westphalia Press
An imprint of Policy Studies Organization
1367 Connecticut Avenue NW
Washington, D.C. 20036
info@ipsonet.org

ISBN: 978-1-63723-663-5

Cover and interior design by Jeffrey Barnes
jbarnesbook.design

Daniel Gutierrez-Sandoval, Executive Director
PSO and Westphalia Press

Updated material and comments on this edition
can be found at the Westphalia Press website:
www.westphaliapress.org

SPACE EDUCATION & STRATEGIC APPLICATIONS

SPACE EDUCATION AND STRATEGIC APPLICATIONS JOURNAL

VOL. 6, NO. 1 • JANUARY 2026

PSO

Letter from the Editors

As we enter 2026, the space domain continues to evolve from a frontier of pure technological conquest into a complex environment requiring sophisticated legal frameworks, sustainable management strategies, and a deeper understanding of human resilience. The challenges we face today are no longer limited to propulsion and orbit; they now encompass the nuances of international law, the preservation of our home planet's ecosystems, and the psychological well-being of the professionals who drive this industry forward.

This issue of the *Space Education and Strategic Applications* (SESA) Journal reflects this expanding scope, bringing together a diverse collection of research that bridges the gap between technical capability and sustainable practice.

The issue opens with a critical examination of the human factor in aerospace environments. As the industry moves beyond merely surviving the harshness of space to establishing long-term presence, the mental health of our workforce is paramount. We present an article advocating for evidence-based approaches to "crafting" personal narratives in aerospace healthcare. This research suggests that by embracing narrative practices, professionals in high-stress environments can foster positive upward spirals, moving from a state of survival to one of genuine thriving.

Shifting focus from the internal human experience to planetary stewardship, this issue also highlights the vital role of space-based technologies in Earth management. We feature a student submission detailing the importance of remote sensing in monitoring plant health and ecosystems. As environmental concerns grow, the ability to utilize airborne and satellite imagery to inform habitat managers and policymakers is becoming increasingly crucial for maintaining healthy, sustainable ecosystems.

On the legal and geopolitical front, this issue tackles the friction between outdated governance and modern commercial realities. With the geopolitical landscape vastly different from when the Outer Space Treaty of 1967 was established, we present a compelling argument for the creation of "Near Space Exclusive Economic Zones." Drawing on maritime law and the United Nations Convention on the Law of the Sea, this article proposes a new legal framework that grants space-faring nations economic rights up to the Kármán line without imposing full sovereignty, aiming to deter conflict and encourage economic development.

Finally, we are proud to present our Student Scholarship Selection, which provides a comprehensive review of the challenges facing a crewed mission to Mars. This article analyzes NASA's three-phase plan—from the ISS to the Proving Ground in cislunar orbit, and finally to the Red Planet—while candidly addressing

doi: 10.18278/sesa.6.1.1

the technological gaps and the increasing significance of commercial contributions given the uncertainties surrounding the Space Launch System.

Against this backdrop of innovation and inquiry, we hope these articles spark robust discussion regarding how we govern, inhabit, and protect the space domain. We express our appreciation to the authors who contributed to this issue, our editorial board for their continued diligence, and our readers for their support.

Sincerely,

Kristen Miller & Gary Deel

Co-Editors in Chief, *SESA* Journal

STUDENT SCHOLARSHIP SUBMISSION

The Importance of Understanding Plant Health and Remote Sensing's Role in Providing the Information for Proper Management

Aaron L. Wood
Department of Space Studies
American Military University

ABSTRACT

The health of the environment starts with the vegetation that resides within the many ecosystems of the planet and the steps necessary to manage their health. Researchers are required to maintain long-term monitoring data that stretches across varying habitats and the different types of plants that make them up. Over the last two to three decades the technologies of remote sensing—the practice of observing and collecting data on an object or region on the Earth's surface—has played a crucial role in assisting habitat managers and policymakers, providing them with necessary information to make informed decisions on how to maintain a healthy and thriving ecosystem. Passive and active sensors, both on airborne and satellite platforms, have obtained a wealth of long-term imagery for these monitoring operations. Without healthy and sustainable plant life, not only will the regional ecosystem and the inhabitants living within it suffer, but so could the entire world. Wetlands, forests, farmland, and grasslands all serve many purposes in the well-being of the environment and the people who depend on them for their resources. This paper outlines these areas of interest and the need for proper management of the health of these plants and their environments.

Keywords: remote sensing, ecosystem, vegetation, satellites

 doi: 10.18278/sesa.6.1.2

La importancia de comprender la salud de las plantas y el papel de la teledetección en el suministro de información para una gestión adecuada

RESUMEN

La salud del medio ambiente comienza con la vegetación que reside en los numerosos ecosistemas del planeta y las medidas necesarias para gestionarla. Los investigadores deben mantener datos de monitoreo a largo plazo que abarcan diversos hábitats y los diferentes tipos de plantas que los componen. Durante las últimas dos o tres décadas, las tecnologías de teledetección —la práctica de observar y recopilar datos sobre un objeto o región de la superficie terrestre— han desempeñado un papel crucial para ayudar a los administradores de hábitats y a los responsables políticos, proporcionándoles la información necesaria para tomar decisiones informadas sobre cómo mantener un ecosistema sano y próspero. Los sensores pasivos y activos, tanto en plataformas aéreas como satelitales, han obtenido una gran cantidad de imágenes a largo plazo para estas operaciones de monitoreo. Sin una vida vegetal sana y sostenible, no solo sufrirá el ecosistema regional y sus habitantes, sino también el mundo entero. Los humedales, los bosques, las tierras de cultivo y los pastizales cumplen múltiples funciones para el bienestar del medio ambiente y de las personas que dependen de ellos para obtener sus recursos. En este artículo se describen estas áreas de interés y la necesidad de una gestión adecuada de la salud de estas plantas y sus entornos.

Palabras clave: teledetección, ecosistema, vegetación, satélites

了解植物健康状况的重要性以及遥感技术在提供正确管理信息方面的作用

摘要

环境健康始于地球上众多生态系统中的植被，以及管理这些植被健康所需的各项措施。研究人员需要维护那些涵盖不同生境及其植物类型的长期监测数据。过去二三十年，遥感技术——即观测和收集地球表面物体或区域数据的技术——在协助生境管理者和政策制定者方面发挥了至关重要的作用，为他们提供了必要的信息，以便就如何维护健康繁荣的生态系统作出明智的决策。被动式和主动式传感器（无论是在机载平台还是卫星平台上）都已获取了大量用于这些监测工作的长期影像。如果没有健康且可持续的植物生命，不仅区域生态系统及其居民会遭受损失，整个世界也可能受到影响。湿地、森林、农田和草原在环境福祉以及依赖它们获取资源的人们的福祉方面都发挥着多种作用。本文概述了这些研究领域，并强调了“妥善管理这些植物及其生长环境健康”一事的重要性。

关键词：遥感，生态系统，植被，卫星

Introduction

All fauna on Earth require oxygen to breathe in order to survive. The plants that grow on this planet provide this oxygen as a byproduct of a process known as photosynthesis, where energy is harvested from sunlight and carbon dioxide is absorbed from the atmosphere by plants and then converted into soluble carbohydrates for the plant's continued growth (Dey et al., 2021 and Shackira et al., 2022). Without this oxygen, life on this planet that draws breath to live would not exist. The importance of maintaining strong, healthy flora stretches beyond the need to take in oxygen to survive. There are a number of other important roles that plants play in this world. The crops that are grown all over the world are consumed as part of an essential diet by humans and other animals, the many different species of trees in the forests provide shelter to the animals that call these forests their home and provide lumber for houses for humans as well, and aquatic vegetation, such as algae, help keep the waters of this planet clean and oxygenated (Campbell & Wynne, 2011). Without all these varying plant species, key processes—such as regulating the global carbon cycle to maintain an equilibrium of the Earth's climate—would cease to exist (Jarvis et al., 2001). Monitoring and maintaining the health of the plant life is essential to the continued health of everything that calls Earth home, including the Earth itself. Scientists and researchers perform tasks of tracking Earth's flora with many different forms of remote sensing, the technology and technique designed to image large geographical areas in a shorter amount of time than more traditional *in situ* forms of data acquisition.

Remote sensing is the gathering of information about an object or geographical area from a distance by use of sensors that detect energy within the electromagnetic spectrum via aircraft or spacecraft (Campbell & Wynne, 2011). This allows for a much larger area of coverage for data collection in a shorter amount of time than direct, physical collection during *in situ* activities. Direct, hands-on collection may involve a single individual or a team of technicians working for extended periods of time, ranging from hours to days to weeks, operating numerous different devices to gather information on plant coverage, species identification, photographs, soil condition measurements, etc. that either an airborne or satellite remote sensing platform could do in minutes to hours. Aircraft are desirable for regional coverage, as they are cheaper to operate than space-based platforms and can be operational and on-site when needed. Aircraft can make multiple passes over the same area, on the same day, in case the first pass is suspected to have insufficient results. Satellites, however, can cover a greater geographical area and can be utilized by a much larger group of researchers, due to their location, orbiting the Earth. As opposed to a team of individuals in a field, one satellite can make observations that can later be analyzed for many different points of concern, the same as those listed for *in situ* collections. This paper will examine the mission objectives and planning, sensor technologies, and vegetation of varying ecosystems, along with direct, contributing factors required for a better understanding of remote sensing applications, such as data processing, soil conditions, and water resources, all of which hold importance to the study of Earth's plant life.

Mission Objective

The study of vegetation has long been of importance to society. As humans expanded across the globe, learning which plants are safe to eat, which ones are most useful for building homes and tools, and what they can tell us about the area, they have established communities and thrived. Plants have been used as indicators to gauge the environmental health of regions, as they can inform individuals of quality lands for agriculture, safe places to live, and the potential for potable water (Kollmann & Fischer, 2003). From the trial-and-error observation methods used centuries ago, the study of plants has evolved to the use of remote sensing systems such as air and spacecraft. The primary goal of vegetation observations and research operations is to better understand the past, current, and future state of the region under study. These studies can range from crop production to disease tracking. Ultimately, observers want to be able to obtain data to support future policy decisions, whether that's the planting of particular crops the next season for, or allowing forestry services to authorize harvesting trees for lumber (Campbell & Wynne, 2011).

Regardless of the operation, proper mission planning is essential for a safe and successful outcome, and the decisions about how to acquire the appropriate data depends on the type of study being conducted. Without the proper data acquisition platform or sensor system, the success of the mission is in jeopardy. What time of the year the observations be taken? Will the observations neccesarily be taken during a specific period, regardless of weather conditions? How large of an area is under study? Will the study require photographs, and if so, will panchromatic be sufficient or will color be needed? These and many other questions must all be considered when planning and preparing for any research project. In this paper, we will be looking at remote sensing observations form aircraft and satellite systems and how they are applied to different situations.

Airborne Platforms

Airborne platforms, such as airplanes, balloons, and kites, have all been used at some time or another for remote sensing of vegetation all over the globe. Whether these observations are just images of an area, or they are intended for a much more scientific purpose, airborne remote sensing plays a vital role in regional vegetation monitoring. The use of aircraft for vegetation studies has shown to be a better option than satellites for smaller, regional studies conducted by either county-level or state-level organizations. The costs associated with the use of a satellite might be out of the budget for localized observations of tree coverage of a state park or the monitoring of crops of agricultural operations (Campbell & Wynne, 2011). The categories of airborne remote sensing platforms can be broken down into two types: manned and unmanned.

Manned airborne platforms, primarily airplanes, have been utilized to capture much higher spatial resolution images—2.5 cm as compared to the 30 cm resolution provided by the Worldview4 satellite—which is the orbiting system with the highest resolution imagery commercially available (Hollings et al., 2018) for surveys and land observations. When satellite observation is not available, yet larger areas still need to be imaged, manned aerial systems have a greater ability to cover these larger areas than unmanned aerial systems (UAS) (Wang et al., 2019). Observing land cover and vegetation from a manned light aircraft allows for the potential of multiple observers scanning the area that is being flown over. UASs only allow for imaging to be taken by the onboard photography system, whereas the personnel inside the aircraft, possibly two in a light aircraft such as a small plane or helicopter, have the ability to scan the land visually and make onboard decisions if another pass may be needed to catch something that may have been missed on the initial pass. There are some benefits to having personnel onboard an airborne observation survey, but there are also many advantages to unmanned surveys as well.

Unmanned operations, carried out by various types of UASs, are fast becoming the platform of choice for aerial remote sensing in vegetation, and other land observation, studies, as compared to manned operations they are generally less expensive and greatly reduce the risk for survey personnel (Wang et al., 2019). Both manned and unmanned aerial imagery platforms have the advantage of being able to customize the sensor system onboard; they are able to have the flights timed when considering the weather; they can measure the time duration between imaging; and they support a faster application of advanced technology (Hollings et al., 2018). Since their introduction to remote sensing, UASs have been applied to the observations of marine fauna (Colfax et al., 2017), as well as animals that live in open areas such as elephants with the use of photographic imagery and thermal infrared sensors for deer and foxes that reside in heavily vegetated areas (Wang et al., 2019), and in agricultural practices (Hunt & Daughtry, 2017). UASs are not only widely being put into use for remote sensing activities, but are also the original remote sensing systems, dating back to 1858 when Tournachon took photographs of Paris from a hot-air balloon and English meteorologist E.D. Archibald utilized kites in 1882. Since that time, three categories of UAS platforms, Medium-Range Endurance to Exo-Stratosferic, close-short-medium-range, and nano-micro-mini have been put into operation, all catering to different functions and conditions. The latter two are more suitable for scientific and commercial research studies of vegetation, as the first is generally operated by the military (Colomina & Molina, 2014).

Satellite Platforms

Earth-observing satellites, once they are in orbit, are a more inexpensive asset for vegetation studies (Wang et al., 2019) than the constant use of airborne platforms. For studies taking place in large areas, such as the prairie grasslands of North America, Landsat, one of the primary satellite systems for land observations, orbits the Earth in a sun-synchronous orbit, traveling from north to south, viewing the sunlit side of the Earth, allowing it to take images of the planet's surface (Guo et al., 2005). Figure 1, below, illustrates the north-to-south orbital track of a sun-synchronous satellite, along with key points, such as the perigee (close point to the Earth), apogee (furthest point from the Earth), ascending and descending nodes (where the satellite crosses the equator on its respective north-south orbit), and the inclination (the angle between the orbital track and the equator at the descending node) (Campbell & Wynne, 2011).

Sun-Synchronous Orbit with Key Points

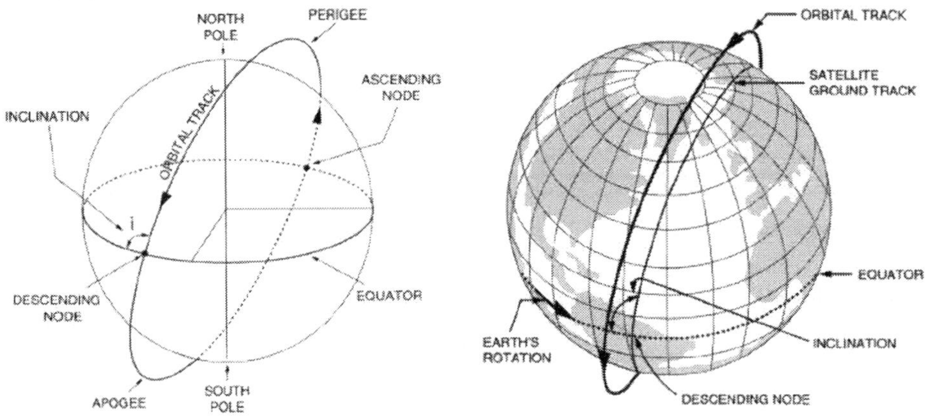

Figure 1. Adopted from (Campbell & Wynne, 2011).

Campbell & Wynne (2011) also point out that for vegetation surveys, photographic imagery is a necessity for an overall understanding of how the plants' growth and coverage progress through the seasons, making the sun-synchronous orbit for remote sensing ideal, allowing satellite systems such as Landsat to observe the same place on the Earth's surface at the same time each day, which is crucial for long-term, continuous studies.

Another advantage of the sun-synchronous orbit, making it desirable for vegetation studies and monitoring, is that it reduces the variations of lighting conditions by synchronizing the orbital track of the satellite to the rotation of the Earth so that it passes over the same ground track at the same time of each successive day (repeat cycle) (Saboori et al., 2014). The *hour angle (h)*, is the angle between the direct solar beam and the point of interest (and respectively, the sensor in orbit), measured with respect to the center of the Earth, and can be calculated with the formula,

$$h = [(GMT - 12.0) * 15] - longitude \quad \text{(Eq. 1)}$$

- GMT = Greenwich Mean Time
- longitude = longitude of the point of interest

With the longitude constantly changing, mission design engineers need to ensure that the satellite orbits the same scene at the same local sun time to maintain an orderly local sun angle (Campbell & Wynne, 2011).

Landsat TM Sensor Parameters and Characteristics

Band	Resolution	Spectral definition	Some applications[a]
1	30 m	Blue-green, 0.45–0.52 µm	Penetration of clear water; bathymetry; mapping of coastal waters; chlorophyll absorption; distinction between coniferous and deciduous vegetation
2	30 m	Green, 0.52–0.60 µm	Records green radiation reflected from healthy vegetation; assesses plant vigor; reflectance from turbid water
3	30 m	Red, 0.63–0.69 µm	Chlorophyll absorption important for plant-type discrimination
4	30 m	Near infrared, 0.76–0.90 µm	Indicator of plant cell structure; biomass; plant vigor; complete absorption by water facilitates delineation of shorelines
5	30 m	Mid infrared, 1.55–1.75 µm	Indicative of vegetation moisture content; soil moisture mapping; differentiating snow from clouds; penetration of thin clouds
6	120 m	Far infrared, 10.4–12.5 µm	Vegetation stress analysis; soil moisture discrimination; thermal mapping; relative brightness temperature; soil moisture; plant heat stress
7	30 m	Mid infrared, 2.08–2.35 µm	Discrimination of rock types; alteration zones for hydrothermal mapping; hydroxyl ion absorption

[a]Sample applications listed here; these are not the only applications.

Table 1. Adopted from (Campbell & Wynne, 2011).

A few of the systems that are in operation for vegetation monitoring are the Landsat Thematic Mapper (TM) and the VEGETATION (VGT) instrument onboard SPOT 4 and SPOT 5. While the TM records in seven spectral bands each with their own applications (Table 1), VGT has a charged-couple device

(CCD) linear array that operates in four spectral bands. VGT's four bands' spectral definitions correspond with bands 1, 3, 4, and 5 respectively (Campbell & Wynne, 2011). All four of these spectral bands are useful for different vegetation scanning, from distinguishing between different plant types to the health of the plants under observation.

Sensor Technology

Regardless of whether the remote sensing platforms are airborne or satellite, there are two different categories of sensor technology: active and passive. The primary difference between the two is that active generates its own energy, and passive detects the energy emitted from the Earth's surface. Each of these two sensor technologies has different advantages and applications when it comes to vegetation studies. The following sections will discuss not only the uses of each of these technologies, but also some of the systems that are in place for each.

Active Microwave Sensors

Active sensors play a vital role in studying vegetation of all different types to determine the growth and health of an ecosystem, from forests to crops. They are especially beneficial on satellite remote sensing platforms. One study has shown that remote sensing methods using Earth observation (EO) satellites equipped with Synthetic Aperture Radar (SAR) are preferred over passive, optical sensors for their ability to penetrate cloud cover, allowing their use regardless of the current weather (Hosseini & Hayne, 2021). This ability allows their use to be scheduled more consistently, as data is able to be obtained from every pass of an area, even if adverse conditions are present, whereas a passive system, using an optical sensor, would not be able to capture images through cloud cover. This can be especially important in times of disaster relief after a hailstorm or hurricane, when analysts would like to know the current conditions of plant life yet the cloud coverage hasn't moved away from the affected area. SAR systems are important for agricultural research as well. One study, conducted by Parag et al. (2024), showed that two SAR systems, RADARSAT-2 and Sentinel-1, were the predominate systems when compared to 13 other SAR sensors that also performed the same function of vegetation biomass studies, and when compared to optical sensor systems, these active sensor systems were found to be more suitable for continuous monitoring as well. Parag et al. (2024) point out that both the RADARSAT-2 and Sentinel-1 SAR satellites offer day and night imaging, multiple polarimetric modes, and a range of incidence angles; however, since it was put into operation in 2014, Sentinel-1 has been utilized more often than RADARSAT-2, due to its shorter revisit time of 12 days, as compared to RADARSAT-2's revisit time of 24 days (cited in Harfenmeister et al., 2021) and its open-access data policy (cited in Khanbbazan et al., 2019).

Post-disaster observation studies are another operation in which active microwave sensors are utilized. SAR satellite images, provided by the International Charter for Space and Major Disasters, have been used to study wind-thrown forests in Sweden, and other European countries, after incredible damage had been caused by devasting storms from 2005 to 2007 (Fransson et al., 2010). While optical imaging can provide high-resolution imaging of effected areas, active radar sensors can provide details that go deeper than the surface cover. Analysts may want to know how these downed trees affect the soil in these areas, and without the ability of radar to penetrate foliage, these readings are only attainable by *in situ* observations.

Passive Optical Sensors

Passive microwave sensors, which are sometimes referred to as optical sensors the electromagnetic radiation energy emitted from the Earth's surface, including vegetation, resulting in panchromatic (black and white; brightness of the radiation within the visible spectrum provides a detailed image without separating the colors), natural-color model (images in the red, green, and blue [RGB]), color infrared model (the blue wavelength band is discarded and a band in the near-infrared [NIR] spectrum is added; green appears as blue; red appears as green; NIR appears as red), and thermal imagery (Campbell & Wynne, 2011). All of these categories of optical imagery serve a purpose in both focused and general-purpose vegetation studies, dependent on the parameters set forth by the research analyst.

The satellite sensors of the Landsat TM and Enhanced Thematic Mapper plus (ETM+), with a 185 km wide swath, its 6 fine resolution channels of 30 meters, a 16-day revisit schedule, and large archive of detailed images dating back decades, provides researchers with a wealth of data to analyze for their particular study, such as the long-term monitoring of the health of the vegetation in four separate national parks of northern Canada, with the addition of the Normalized Difference Vegetation Index (NDVI) (Fraser et al., 2011). One of the most common vegetation indices, the NDVI, when used with Landsat TM data, is calculated as shown in the below equation (Campbell & Wynne, 2011).

$$NDVI = \frac{NIR-R}{NIR+R} = \frac{TM4-TM3}{TM4+TM3} \qquad \text{(Eq. 2)}$$

Passive, optical satellite sensors are desirable because they are cheaper than eddy-covariance measurement methods, which are measurements taken of the carbon, water, and heat flow between vegetation and the atmosphere (Liang et al., 2012); however, they vary greatly in illumination, weather conditions, and sensor noise, but these variations can be minimized by the use of optical sensor systems near the ground (Eklundh et al., 2011). Airborne sensor systems, installed on aircraft, can fly below the cloud cover and is able to obtain the same sun angle across multiple scans of the vegetation under study. The Compact Airborne Spectrographic Imager-1500 is a hyperspectral imager that has been used, in conjunction with the relative spectral responses of current and former multispectral satellite sensors, to analyze such things as submerged aquatic vegetation (Rowan et al., 2021). Passive sensor systems like these, both airborne and space-based, are a vital asset in the surveying of vegetation of all ecosystems, and a necessary resource in the monitoring of plant life around the world.

Plant Categories Observed

The study of vegetation on Earth via remote sensing data has seen tremendous growth in the last two to three decades, due in part to vast quantities of rapidly growing and freely accessible data from both airborne and satellite optical sensors (Kulawardhana et al., 2017). These sensors operate within the visible spectrum (400 to 700 nm) (Pinter Jr. et al., 2003), utilizing electromagnetic energy emitted from the plants on Earth's surface. The green color that is associated with healthy vegetation is due to the way that chlorophyll pigment in plants absorb energy in the red and blue portions (0.6 to 0.7 μm and 0.4 to 0.5 μm, respectively) of the electromagnetic spectrum and how the mesophyll structures reflect the energy in the green and even higher in the near-infrared portions (0.5 to 0.6 μm and 0.72 to 1.30 μm, respectively) (Wu et al., 2023; Singh & Vyas, 2022; Basso et al., 2004; Campbell & Wynne, 2011), allowing for the employment of the NDVI (Eq. 2) to calculate an ecosystem's capability to convert the energy from the Sun into biomass (Wu et al., 2023, cited in An et al., 2013). Being able to capture detailed images of different vegetation and their environments is crucial to the researcher's and policymaker's ability to make informed decisions on the health and productivity of the plant life under observation. Aquatic vegetation, trees in forests, varying crops in a farmer's fields, and grasses in the prairies can inform analysts about the health of the ecological processes in those ecosystems.

Aquatic Vegetation

About 70 percent of the Earth's surface is water, both fresh and marine, and is a major resource for humans and animals alike, providing potable water, food resources, and modes of travel. As with plants of any other ecosystem, the local, submerged aquatic vegetation is essential to the ecological health of bodies of water, as it supplies the local fauna with food and shelter, and all around improves the water quality (Rowan et al., 2022, cited in Hestir et al., 2016). Singh and Vyas (2022) state that "one of the most effective ways to monitor riverine environments, including lakes, channels, reservoirs, floodplains, and wetlands, is to use remote sensing and geographic information systems." The utilization of these two tools provides a crucial source of information mapping that analysts use to study an ecosystem's vegetative health.

Optical remote sensing appears to be the desired method for studying and monitoring aquatic vegetation, including submerged plants (Rowan et al., 2021), as there is a large inventory of ongoing remote sensing imagery in which to apply derived vegetation indices in order to estimate the biomass in various wetlands (Wu et al., 2023). The separation of water and vegetation is distinguished in near-infrared and medium-infrared wavelengths (0.72 to 1.30 and 1.30 to 3.00 μm, respectively) (Campbell & Wynne, 2011), as water primarily absorbs this energy and vegetation reflects a greater amount (Singh & Vyas, 2022). Both Wu et al. (2023) and Singh & Vyas (2022) utilized the Landsat constellation's Thematic Mapper, Enhanced Thematic Mapper Plus, and Operational Land Imager sensors to obtain imagery data, and Wu et al. (2023) also applied Google Earth Engine image services to create their monthly image composites for their studies. Large composite imagery collections, such as those employed in these two studies, are essential for long-term monitoring and accurate vegetation studies.

Crops

A healthy agricultural system is vital for food security for a nation, as well as for a vibrant and productive ecosystem and surrounding environments. Strong, healthy crops can provide not only food resources for humans, but also food and shelter for wildlife and protection from soil erosion that can have a positive impact on the cleanliness of surrounding waterways. The health of agricultural vegetation and the production of crops have direct effects on the health and sustainability of a large part of the globe (Basso et al., 2004), and since as early as 1929, aerial photographs have been used to map crop production and the environment associated with agricultural practices (Basso et al., 2004, cited in Bauer, 1975). Agricultural operations do not plant, grow, and harvest just one type of crop, and there is likely more than one type of crop grown in adjacent fields within a farm. Remote sensing data can provide information on each of those crops, based on a strong understanding of the local vegetation by the analyst studying images taken to monitor crop production and growing processes.

The knowledge of the growth stages of specific crops in a particular region (i.e., the Midwest of the United States) allows farmers and agronomists to develop a crop calendar, which is a timeline that outlines the growth stages of a crop, such as winter wheat (Table 2) and corn (Table 3) in Kansas, to a calendar month. With this knowledge, remote sensing imagery will allow analysts to observe the growth and health of different crops in a single field or image (Campbell & Wynne, 2011).

Management of crop growth, health, and production is important in driving economic and environmental decisions and has been improved upon over the years by utilizing remote sensing imagery of farm crop fields. Ideally, collection of multispectral imagery should be taken early in the day, at the beginning of the growth stage, as the sun angle at this time of the day allows for greater detection of plants in low-leaf stages (Pinter Jr. et al., 2003). Starting at the beginning of the growth stage and continuing through the growing season, surveys of the same crops in the same region can provide the necessary

Crop Calendar for Winter Wheat

September	Plowing and planting
October	Emergence and early growth
November	
December	
January	Dormancy
February	
March	
April	Growth resumes
May	
June	Heading
July	Harvest
August	

Table 2. Adopted from (Campbell & Wynne, 2011).

Crop Calendar for Corn

April	Plowing and planting
May	Emergence and early growth
June	
July	Tasseling
August	Maturity and harvest
September	

Table 3. Adopted from (Campbell & Wynne, 2011).

data to farmers about when to plant and when to harvest. Long-term monitoring of crop production across multiple seasons can also provide important data on the health of the crop's growth potential from year to year and allow analysts to monitor environmental impacts potentially resulting from the region's agriculture. Further studying of agricultural practices as a whole, not just regionally, is important to maintaining a healthy balance between human interaction (farming the land) and ecosystem responses (negative, positive, or neutral) to those operations.

Forests

Forests are one of the most important ecosystems on the planet. They provide a large amount of oxygen, which is necessary for life, and also act as a major component in absorbing and recycling carbon from the atmosphere. Remote sensing plays an important role in monitoring the health of forests and supplies important data to policymakers, allowing for informed decisions on environmental and industrial practices concerning the use of forest resources. Studying forests is important to environmental health monitoring, but also forestry productivity for timber harvesting operations, and understanding tree diversity may lead to an increased ability for forest management for the timber industry (Mori et al., 2017). Fassnacht et al. (2016) provided a review of over 100 studies that have shown the importance of remote sensing in the classification of different species of trees as it relates to the management of these forests. Spatial information about the tree species over a large area is essential to the understanding of the ecology of forests and the roles that they play in community dynamics within the ecosystem (Fassnacht et al., 2016, cited in Chambers et al., 2013). Species classifications for a number of these studies were obtained by hyperspectral spectroscopy and numerous multispectral resolution systems, such as Landsat and Hyperion (moderate), SPOT and RapidEye (high), and IKONOS (very high), where most of these passive sensors were coupled with active sensor systems (Fassnacht et al., 2016).

Forest fires are a major concern in certain regions of the world, including the southwestern and western United States. Each year they cause tremendous damage to habitats and populated areas, destroying homes and sometimes entire cities (Barmpoutis et al., 2020). One of the biggest concerns when dealing with a forest fire is the safety of those individuals on the ground. Advances in remote sensing technologies, especially in the field of unmanned aerial vehicles (UAVs), and the utilization of "ultra-high resolution optical or infrared charge-coupled (CCD) cameras" have provided a safe means to observe and track, with a greater level of safety and accuracy, the progression of the fire from above (Barmpoutis et al., 2020). While the specially equipped UAVs can survey an active forest fire, preemptive monitoring of the forest floor may assist forest managers in carrying out targeted risk mitigation activities to help prevent these fires from either starting or spreading too rapidly to respond to in a timely manner. Photographs from above the forest may not reveal the debris on the forest floor, which will act as fuel to start and spread fires faster. Lidar and radar sensors provide their own energy source for imaging surfaces of Earth and have the ability to penetrate through crown coverage of trees to observe the forest floor for these types of surveys. SLICER (Scanning Lidar Imager of Canopies by Echo Response) is a lidar (light detection and ranging) sensor system, a form of active sensor, that can be applied to studying the structure of vegetation canopies of trees, which is vital to understanding the physical structure of a forest and has become a proven inventory tool for forest management operations (Campbell & Wynne, 2011). By studying living trees and other vegetation in forests, not only can forest managers determine the health and productivity of a forest for timber harvesting operations, but can also take appropriate actions to maintain and clean up debris-ridden areas that are at potential risk of fire.

Grasslands and Prairies

Grass and prairie lands are the most expansive terrestrial ecosystems in the entire world, covering approximately 30% (Bengtsson et al., 2019) of the planet's landmass, and are a critical food source for various types of livestock (Ali et al., 2016). As with all other ecosystems, the quality and health of the grasses in prairie lands is an indicator of the health and productivity of the local ecology.

Remote sensing, with a focus on grassland management, is utilized for determining and classifying different types of grasses, biomass estimations for the assessment of livestock grazing capacity, pasture and vegetation quality, and vegetation growth rates (Ali et al., 2016).

Grasslands play an important role in the local ecosystems as they affect conditions such as water supply, climate regulation, erosion prevention, livestock for quantity and quality, and crop production in croplands (Bengtsson et al., 2019). Unlike other ecosystems, such as forests and wetlands, the vegetation of the prairie lands is not overbearing to the point of obscuring other plants that analysts wish to study. Landsat's optical imaging sensors, especially, are the primary systems used to study grasslands and monitor their health and productivity, because lidar systems (active microwave sensors) are not suitable for studying and surveying grasslands (Kulawardhana et al., 2017). These systems may not be able to detect centimeter-level differences between different grasses and may not be able to distinguish between vegetation and ground surfaces (Kulawardhana et al., 2017, cited in Bowen & Waltermire, 2002).

Critical Thinking

The application of both passive and active remote sensing sensors for the long-term monitoring of plants across varying ecosystems has been extensively covered in professional research. Depending on the plant life being surveyed, and the studies being conducted, researchers make the decision on which sensor technology to employ, and whether airborne or space-based is more applicable. With the continual advancements in remote sensing technologies and the increase in environmental awareness, planetary sustainability, safety, the well-being of society, and research into the vegetation of ecosystems will continue to increase as well. Data provided from multiple remote sensing platforms, passive and active, and airborne and space-based, will continue to play an important role in monitoring and managing the health of Earth's vegetation. As vast as the planet is, and expansive areas that can be extremely remote and difficult for individuals to physically access, without the capabilities of remote sensing systems, those regions may go unmonitored and managed, which could lead to devasting results.

Conclusion

While there are many different types of ecosystems in the world, all with their own primary types of vegetation, the one thing that is consistent across them all is that all successful monitoring and management of these plants is needed to ensure a healthy ecosystem. Plants in all environments play roles in providing clean and habitable conditions by providing oxygen for growth, shelter for fauna including timber that humans use in their construction of homes, and as a food source. Remote sensing, both active and passive, plays an essential role in proper management of these plants and their regions. With the vastness of these habitats, eddy-covariance and in situ measurements cannot supply researchers and analysts with the proper level of detail or the quantity of data that is needed to perform proper management practices. As it stands today, continued, long-term data collection and monitoring of the planet's vegetation, both great and small, is vital to the continued health of the ecosystems and the sustainability of their production to support human activities.

References

Barmpoutis, P., Papaioannou, P., Dimitropoulos, K., & Grammalidis, N. (2020). A review on early forest fire detection systems using optical remote sensing. *Sensors, 20*. https://doi.org/10.3390/s20226442

Basso, B., Cammarano, D., & Vita, P. D. (2004). Remotely sensed vegetation indices: Theory and applications for crop management. *Rivista Italiana Di Agrometeorologia, 1*(5), 36–53.

Bengtsson, J., Bullock, J. M., Egoh, B., Everson, C., Everson, T., O'Connor, T., O'Farrell, P. J., Smith, H. G., & Lindborg, R. (2019). Grasslands—more important for ecosystem services than you might think. *Ecosphere, 10*(2). https://doi.org/10.1002/ecs2.2582

Campbell, J. B., & Wynne, R. H. (2011). *Introduction to Remote Sensing* (5th ed.). Guilford Press.

Colefax, A. P., Butcher, P. A., & Kelaher, B. P. (2017). The potential for Unmanned Aerial Vehicles (UAV) to conduct marine fauna surveys in place of manned aircraft. *ICES Journal of Marine Science, 75*(1), 1–8. https://doi.org/10.1093/icesjms/fsx100

Colomina, I., & Molina, P. (2014). Unmanned aerial systems for photogrammetry and remote sensing: A review. *ISPRS Journal of Photogrammetry and Remote Sensing, 92*, 79–97. https://doi.org/10.1016/j.isprsjprs.2014.02.013

Dey, P., Datta, D., Pattnaik, D., Dash, D., Saha, D., Panda, D., Bhatta, B. B., Parida, S., Mishra,U. N., Chauhan, J., Pandey, H., & Singhal, R. K. (2021). Chapter 5 - Physiological, biochemical, and molecular adaptation mechanisms of photosynthesis and respiration under challenging environments. In *Plant Perspectives to Global Climate Changes: Developing Climate-Resilient Plants*. Essay, Academic Press.

Eklundh, L., Jin, H., Schubert, P., Guzinski, R., & Heliasz, M. (2011). An optical sensor network for vegetation phenology monitoring and satellite data calibration. *Sensors, 11*, 7678–7709. https://doi.org/10.3390/s110807678

Fassnacht, F. E., Latifi, H., Stereńczak, K., Modzelewska, A., Lefsky, M., Waser, L. T., Straub, C., & Ghosh, A. (2016). Review of studies on tree species classification from remotely sensed data. *Remote Sensing of Environment, 186*, 64–87. https://doi.org/10.1016/j.rse.2016.08.013

Fransson, J. E. S., Pantze, A., Eriksson, L. E. B., Soja, M. J., & Santoro, M. (2010). Mapping of wind-thrown forests using satellite SAR images. *2010 IEEE International Geoscience and Remote Sensing Symposium*. https://doi.org/10.1109/igarss.2010.5654183

Fraser, R. H., Olthof, I., Carrière, M., Deschamps, A., & Pouliot, D. (2011). Detecting long-term changes to vegetation in northern Canada using the Landsat Satellite Image Archive. *Environmental Research Letters, 6*. https://doi.org/10.1088/1748-9326/6/4/045502

Guo, X., Zhang, C., Wilmshurst, J. F., & Sissons, R. (2005). Monitoring grassland health with remote sensing approaches. *Prairie Perspectives, 8*, 11–22.

Hollings, T., Burgman, M., van Andel, M., Gilbert, M., Robinson, T., & Robinson, A. (2018). How do you find the green sheep? A critical review of the use of remotely sensed imagery to detect and count animals. *Methods in Ecology and Evolution, 9*, 881–892. https://doi.org/10.1111/2041-210x.12973

Hosseini, M., & Haynes, K. (2021, October 25). *Monitoring soil moisture over agricultural fields using synthetic aperture radar*. Earth data for informed agricultural decisions. https://nasaharvest.org/news/monitoring-soil-moisture-over-agricultural-fields-using-synthetic-aperture-radar

Hunt, E. R., & Daughtry, C. S. (2017). What good are unmanned aircraft systems for agricultural remote sensing and precision agriculture? *International Journal of Remote Sensing, 39*, 5345–5376. https://doi.org/10.1080/01431161.2017.1410300

Jarvis, P. G., Saugier, B., & Schulze, E.-D. (2001). Productivity of boreal forests. *Terrestrial Global Productivity*, 211–244. https://doi.org/10.1016/b978-012505290-0/50011-9

Kollmann, J., & Fischer, A. (2003). Vegetation as indicator for habitat quality. *Basic and Applied Ecology, 4*(6), 489–491. https://doi.org/10.1078/1439-1791-00209

Kulawardhana, R. W., Popescu, S. C., & Feagin, R. A. (2017). Airborne lidar remote sensing applications in non-forested short stature environments: A review. *Annals of Forest Research, 60*(1), 173–196. https://doi.org/10.15287/afr.2016.719

Liang, S., Li, X., & Wang, J. (2012). Vegetation production in terrestrial ecosystems. *Advanced Remote Sensing* (pp. 501–531). essay, Academic Press. Retrieved 2024, from https://www.sciencedirect.com/science/article/pii/B9780123859549000162.

Mori, A. S., Lertzman, K. P., & Gustafsson, L. (2017). Biodiversity and ecosystem services in forest ecosystems: A research agenda for Applied Forest Ecology. *Journal of Applied Ecology, 54*, 12–27. https://doi.org/10.1111/1365-2664.12669

Parag, M., Lottering, R., Peerbhay, K., Agjee, N., & Poona, N. (2024). The use of synthetic aperture radar technology for crop biomass monitoring: A systematic review. *Remote Sensing Applications: Society and Environment, 33.* https://doi.org/10.1016/j.rsase.2023.101107

Pinter, Jr., P. J., Hatfield, J. L., Schepers, J. S., Barnes, E. M., Moran, M. S., Daughtry, C. S. T., & Upchurch, D. R. (2003). Remote sensing for crop management. *Photogrammetric Engineering and Remote Sensing, 69*(6), 647–664. https://doi.org/10.14358/pers.69.6.647

Rowan, G. S., Kalacska, M., Inamdar, D., Arroyo-Mora, J. P., & Soffer, R. (2021). Multi-scale spectral separability of submerged aquatic vegetation species in a freshwater ecosystem. *Frontiers in Environmental Science, 9.* https://doi.org/10.3389/fenvs.2021.760372

Saboori, B., Bidgoli, A. M., & Saboori, B. (2014). Multi-objective optimization in repeating sun-synchronous orbits design for remote-sensing satellites. *Journal of Aerospace Engineering, 27.* https://doi.org/10.1061/(asce)as.1943-5525.0000340

Shackira, A. M., Sarath, N. G., Aswathi, K. P., Pardha-Saradhi, P., & Puthur, J. T. (2022). Green seed photosynthesis: What is it? What do we know about it? Where to go? *Plant Physiology Reports, 27*(2). https://doi.org/10.1007/s40502-022-00695-4

Singh, A., & Vyas, V. (2022). A review on remote sensing application in river ecosystem evaluation. *Spatial Information Research.* https://doi.org/10.1007/s41324-022-00470-5

Wang, D., Shao, Q., & Yue, H. (2019). Surveying wild animals from satellites, manned aircraft and Unmanned Aerial Systems (UAS): A Review. *Remote Sensing, 11.* https://doi.org/10.3390/rs11111308

Wu, W., Grimes, E., & Suir, G. (2023). Impact of freshwater diversions on vegetation in coastal wetlands based on remote sensing derived vegetation index. *Frontiers in Marine Science*. https://doi.org/10.3389/fmars.2023.1202300

STUDENT SCHOLARSHIP SUBMISSION

Medical, Technological, and Ethical Challenges for the Initial Expedition to Mars

Laura Clark

The American Public University System

ABSTRACT

For more than a decade, the National Aeronautics and Space Administration (NASA) has been actively pursuing the goal of landing humans on Mars. This spectacular mission has been inspired by the search for extraterrestrial life, advancing planetology, and the potential for human habitation. Since landing a man on the moon in 1969, there has been enormous pride associated with space travel in American society. Human spaceflight and exploration should continue to be cultivated in its citizens, government, academic pursuits and commercial endeavors.

NASA's three-phase plan for landing humans on Mars effectively addresses the complexities of deep space exploration by leveraging past lessons and fostering collaborations. Their iterative Moon to Mars Architecture ensures continuous technological advancements, positioning them well to achieve their ambitious goal. Given the current uncertainties of Space Launch System (SLS) viability, however, the availability and affordability of commercial contributions is of increasing significance. This paper explores the medical, technological, and ethical challenges associated with the initial human mission from a lunar outpost to Mars. Additionally, it examines the critical requirements for successfully initiating research and development activities on the Martian surface.

Keywords: NASA, Mars, medical, ethical, Outer space, space travel, technology

doi: 10.18278/sesa.6.1.3

Desafíos médicos, tecnológicos y éticos para la expedición inicial a Marte

Resumen

Durante más de una década, la Administración Nacional de Aeronáutica y del Espacio (NASA) ha perseguido activamente el objetivo de llevar humanos a Marte. Esta espectacular misión se ha inspirado en la búsqueda de vida extraterrestre, el avance de la planetología y el potencial de asentamiento humano. Desde que el hombre pisó la Luna en 1969, la sociedad estadounidense ha sentido un enorme orgullo por los viajes espaciales. Los vuelos espaciales tripulados y la exploración deben seguir siendo un objetivo prioritario para la ciudadanía, el gobierno, las actividades académicas y las iniciativas comerciales.

El plan de tres fases de la NASA para el aterrizaje humano en Marte aborda eficazmente las complejidades de la exploración del espacio profundo, aprovechando las lecciones aprendidas y fomentando la colaboración. Su arquitectura iterativa de la Luna a Marte garantiza avances tecnológicos continuos, lo que la posiciona bien para alcanzar su ambicioso objetivo. Sin embargo, dada la incertidumbre actual sobre la viabilidad del Sistema de Lanzamiento Espacial (SLS), la disponibilidad y asequibilidad de contribuciones comerciales cobra cada vez mayor importancia. Este artículo explora los desafíos médicos, tecnológicos y éticos asociados con la misión humana inicial desde un puesto avanzado lunar a Marte. Además, examina los requisitos críticos para iniciar con éxito las actividades de investigación y desarrollo en la superficie marciana.

Palabras clave: NASA, Marte, medicina, ética, espacio exterior, viajes espaciales, tecnología

首次火星探测任务面临的医学、技术和伦理挑战

摘要

十多年来，美国国家航空航天局(NASA)一直积极致力于实现载人登陆火星的目标。这项宏伟计划源于对地外生命的探索、行星学的发展以及人类在火星上居住的潜力。自1969年人类成功登月以来，太空旅行一直是美国社会引以为豪的事业。载人航天和探索应该继续在美国公民、政府、学术界和商业领域得到大力发展。

NASA的载人火星登陆三阶段计划通过借鉴以往经验和促进合作，有效应对了深空探索的复杂性。其迭代式的“月球到火星架构”确保了技术的持续进步，使其能够更好地实现这一宏伟目标。然而，鉴于目前太空发射系统(SLS)可行性的不确定性，商业贡献的可用性和可负担性变得日益重要。本文探讨了从月球前哨站到火星的首次载人任务所面临的医学、技术和伦理挑战。此外，本文还探讨了“在火星表面成功开展研发活动”一事的关键要求。

关键词：美国国家航空航天局(NASA)，火星，医学，伦理，外太空，太空旅行，技术

Introduction

Collaboration with international and U.S. companies will enable NASA to push past the lunar outpost and onto Mars. This exploration to Mars is not the terminus, but an agent to permanent human deep space occupation. President Obama formally directed the agency to reach for human exploration of the Red Planet by the 2030s (National Aeronautics and Space Administration Authorization Act of 2010). Multiple collaborative partnerships have enabled success in the lunar outpost thus far: communicating a dedication to advancing scientific discoveries and commencement of the lunar economy. The capital investments in technology enable 52 crew members to reside on the Moon for two 5-year rotations. But this trial is intended to enable deep space exploration, and substantial investments are necessary to overcome the medical, technological, and ethical challenges faced today.

This paper will first examine the medical challenges that require further research. Space radiation is still the most formidable obstacle to human exploration. The omnipotence of ionized radiation from Solar Particle Events (SPE) and Galactic Cosmic Radiation (GCR) and the risks posed to biological systems are pertinent topics of study in the space community (Montesinos et al., 2021). Limitations of current shielding technology and nutritional understanding of immunity methods require further studies. Continually, there is technology needed to sustain humans on Mars for the intended 6 to 18 months stay. Life support systems and habitats require further development for this mission to be successful. The multi-faceted robotic deployment to the Red Planet ahead of the explorers will aid greatly in habitat manufacturing (Weisben et al., 1997). Finally, there are policies and procedures that need developing while ensuring adherence to the highest ethical standards. The realization of this daring journey will strengthen America's leadership both in space and back on Earth.

Methods

NASA's *Houston We Have a Podcast* Episode "Moon to Mars: Reviewing the Architecture" (Cheshier, 2024) was used to gather background data on the Architecture Concept review and efforts to take humans into deep space. APU's Tefry Library was utilized to find peer-reviewed, English references. Some of the MeSH terms used to search were spacecraft medical challenges, nanocomposites material space benefits, "US Space Force" ethics, NASA Mars architecture, and radiation countermeasures. Nicogossian et al. (2016) textbook and week 1-8 lesson plans from SPST619 were used for background knowledge and linked references explored. The works referenced were all published after 1997, but the majority from 2010s or more recent. Fair evidence came from two textbooks by McCoy and Kahn (2014) and Nicogossian et al (2016).

Results

From the combined extensive searches, over 14,000 papers were returned. Of this, 37 abstracts were read and filtered to inform this report. 14 peer-reviewed journal articles, 2 legal documents, 1 government publication, 2 reports, 4 .gov websites, 1 NASA podcast, 2 textbooks and 8 APU course SPST619 lesson plans are included in the references section. Life journal has a strong reputation and was awarded an impact factor of 3.2 from the MDPI in 2024. The summary of types of evidence can be seen below in Table 1.

Table 1. *Research and Citation Evaluation Summary*

Author/Date	Publication Type	Remarks	Strength of Evidence
Miranda et al., 2023	Peer reviewed journal article	Evaluates anecdotal evidence regarding multiple lunar explorer's experiences with dust and their issues. Primary and secondary effects from lunar dust are investigated and compared to Earth peer-reviewed studies.	Strong
Paul et al., 2022	Peer reviewed journal article	The first experiment to use samples from Apollo 11, 12, and 17, to show that the terrestrial model plant Arabidopsis thaliana germinates and grows in diverse lunar regolith. Further studies are required.	Strong

Weisbin et al., 1997	Peer reviewed journal article	A review of the role of robotics in space exploration for the four core applications areas.	Good
Stough & Nicogossian, 2006	Report	Medical policy discussion. Thesis from George Mason University.	Fair
Tanaka et al., 2017	Peer reviewed journal article	Microgravity effects on human subjects aboard ISS. Deconditioning and countermeasure suggestions.	Good
Rittweger et al., 2018	Peer reviewed journal article	Analysis of the Sarcolab pilot study into muscle adaptation of long-term spaceflight.	Good
Baiocco et al., 2020	Peer reviewed journal article	PERSEO ISS experiment with a water-filled torso garment to protect against radiation.	Strong
Montesinos et al., 2021	Peer reviewed journal article	Review of Space radiation countermeasures in planetary exploration applications.	Good
Basner et al., 2014	Peer reviewed journal article	Summary of the 520 day simulated Mars mission and the psychological considerations of isolation/confinement.	Good
Kim et al., 2024	Peer reviewed journal article	Review of astronaut personality traits which could mitigate crew interaction risks on a Mars mission.	Good
Bell, 2007	Peer reviewed journal article	Meta analysis of Deep-Level Composition Variables as Predictors of Team Performance, focused on psychology	Strong
NASA, 2015	Government publication	Laid out the strategies and long-term plan to send humans to Mars.	Fair

Pernigoni et al., 2024	Peer reviewed journal article	An experiment with self-healing polymers and nanocomposites where they examined the radiation shielding properties and compared them to typical metallic spacecraft materials.	Strong
Krichevsky, 2021	Peer reviewed journal article	Analysis of the development of artificial gravity for space biosphere environmental management. Introduces philosophical arguments for AG.	Strong

Note. This table shows the relevant sources used to inform this discussion. A summary of their methods is provided. The strength of evidence is based on sample size, research methods, limitations and inconsistencies in the cited works.

Discussion

NASA's three-phase plan for landing humans on Mars involves learning from past missions and collaborating with international, academic, and commercial partners. The first phase, Earth Reliant, included over two decades of experiments aboard the ISS and in isolated environments to prepare for lunar missions. This was followed by the Proving Ground phase, which focused on stable cislunar orbit and exploratory missions that confirmed capabilities for Mars. The final phase, Earth Independent, is when humans will land on Mars, building on the groundwork laid in previous stages.

Key challenges identified by NASA include transportation, working in space, and maintaining astronaut health. Significant technological innovations are necessary to address these demands, as outlined in the Moon to Mars Architecture blueprint. The agency conducts annual Architecture Concept Reviews to assess mission status and gather feedback, producing an updated *Architecture Definition Document* to identify technological gaps that academia and industry can help fill. NASA emphasizes the importance of partnerships and global collaboration in this endeavor.

Medical Considerations

Human interplanetary travel and subsequent habitation on Mars present a wide array of health challenges. The harsh and unforgiving environment of space is characterized by dangers such as microgravity, elevated radiation levels, ICEs, and an absence of atmosphere.

Prolonged exposure to microgravity is associated with health issues such as cardiovascular complications, bone density loss, muscular atrophy, and vestibular disorders (Montesinos et al., 2021; Nicogossian et al., 2016; Rittweger et al., 2018; Tanaka et al., 2016). Altered hydrostatic pressure in microgravity reduces heart stroke volume, impacting cardiovascular health (Tanaka et al., 2016). Rittweger et al. (2018) noted that changes in muscle architecture contribute to muscle weakness. The Sarcolab pilot study showed that astronaut B, who followed a stricter exercise regimen, maintained better muscle strength than astronaut A. Nutritional supplements like selenium and pharmacological interventions may help mitigate some negative health effects of interplanetary travel (Montesinos et al., 2021). The 2015 NASA Twins Study revealed mitochondrial dysregulation as a biological response to extended spaceflight (Krichevsky, 2021). To address microgravity's adverse effects, designing a spacecraft with artificial gravity is a key objective to be explored in the technology section.

Radiation is the prime health risk in spaceflight, with ionizing radiation being far more harmful than what humans are adapted to on Earth (non-ionizing). Ongoing research aims to find nutritional and pharmacological ways to reduce radiation damage. Non-invasive radiation dosimeters will be a mandatory part of the space traveler's uniform. Crew selection is vital, as genetic and biological predispositions yield varying radiation tolerances (Bourguignon et al., 2005; Nicogossian et al., 2016). While lowering radiation exposure limits can reduce costs, further ISS research is needed to confirm these limits. Furthermore, studies must differentiate between continuous and episodic exposure, as well as between high-energy and low-energy particles, which can have opposing effects on biological systems. NASA currently prioritizes astronaut safety through its ALARA (As-Low-As-Reasonably-Achievable) policy. Investigations into anti-radiation medications and genetic vaccines are ongoing, and NASA has used multi-vitamins and minerals in past missions to help mitigate negative radiation effects (Montesinos et al., 2021).

Mental and behavioral health is crucial for the Mars mission, as crew selection, training, and individual resilience will significantly impact success and safety (Kim et al., 2024). The Mars Society and NASA have invested decades in human performance data and ergonomic design, but a 520-day simulated Mars mission revealed that conflicts with mission control were five times more frequent than those among crew members (Basner et al., 2014). Effective communication and conflict resolution strategies will be essential, especially since the Mars team must maintain contact with the lunar outpost, which will serve as a liaison to Earth. Given the inevitable delays in communication, stable team dynamics are vital. Research conducted by Bell (2007) underscores the importance of morale in high-stakes environments. A sustainable work-rest cycle positively influences morale, which will be essential for the success of the approximately 2.5-year-long mission. Psychological preparedness is a critical component of medical readiness.

Technological Considerations

Technology will play a crucial role in the journey from the moon to Mars and back, particularly through the implementation of artificial gravity to offset the negative effects of microgravity on the human body. Countermeasures to microgravity and mandatory daily 2 hours exercise are only partially effective. Artificial gravity is costly but essential for long-duration flights (6-9 months) to ensure astronaut safety and health (Krichevsky, 2021). Research into reliable systems using centrifugal force or high-frequency vibrations is needed, while Mars's more favorable gravity ($\approx$0.38 g) allows for the exclusion of artificial gravity in Martian outposts.

To combat radiation exposure, technological solutions like space suits and radiation shielding are vital (Nicogossian et al., 2016, pp. 43-46). Space suits are necessary for extravehicular activities (EVA), where decompression sickness is a risk. Additionally, the use of self-healing nanocomposites may increase durability and safety while simultaneously reducing weight (Pernigoni et al., 2024). Administrative measures such as aligning launch plans with solar cycle minimums is a low-cost way to decrease total crew radiation exposure.

Radiation protection methods can be categorized broadly into personal and vehicle protection. Personal protection examples include water-filled wearable suits for maneuverability, while vehicle protection involves shielding materials like water or hydrogen (Biacco et al., 2020). However, the Bremsstrahlung effect, which can increase radiation by 30%, must be considered in shielding design (Nicogossian, 2016). Bremsstrahlung translates to “braking radiation” in German and occurs when a high energy electron interacts with an atom's nucleus (i.e., impacts the metallic shield of spacecraft) result in the emission of electromagnetic radiation. Effective radiation shielding requires balancing time, cost, weight, and effectiveness. A hybrid approach of personal and vehicular protection is desirable.

Upon landing on Mars, robots will have prepared an underground habitat, minimizing manual labor for astronauts. Further funding is required to advance the artificial intelligence (AI) capabilities and energy storage of the robotic cadre project managers (Weisben et al., 1997). This habitat will offer radiation and dust protection, featuring four double airlock exits, dormitories, labs, a communal space, and a mess hall. Nuclear power will supply lighting and electrical systems, supplemented by solar panels, with additional funding needed for robotic maintenance to prevent dust-related issues, as seen with NASA's InSight rover in 2022.

Paul et al. (2022) propose using Mars to initiate in situ growth through a regolith matrix, while lunar regolith experiments have shown limited plant growth. Hydroponics will be prioritized on Mars, but establishing a research site for terrestrial agriculture is essential. Plants are crucial for long-term space habitation, as they provide food and oxygen, recycle water, and remove carbon dioxide.

Lunar dust poses significant challenges due to its abrasive and sharp nature. As humans are established on the moon with a semi-permanent presence, integrated dust mitigation strategies are vital for sustainable lunar architecture. Although Martian dust differs from lunar dust, advancements in understanding lunar dust toxicity and mitigation strategies can inform future operations on Mars (Miranda et al., 2023).

Ethical Considerations

The journey from the Moon to Mars raises significant ethical considerations, particularly regarding the governance and utilization of outer space. The Outer Space Treaty of 1967, like the Antarctic Treaty, established a framework for international cooperation in space. But competing political, military, commercial, and scientific interests complicate its application. Just as disregard for maritime agreements can harm shared oceans, violations of space treaties can have far-reaching consequences (*The American Journal of International Law*, 2020).

Leadership and crew composition will play a critical role in addressing these ethical challenges. While decisions will primarily be made democratically, a single mission commander will take control in emergencies. Medical policy formulation and advanced directives will be a particular notable ethical friction point. Informed consent and equitable access to medical treatment for the 10-person exploration crew to Mars will be mandatory. A code of conduct could be implemented for the travelers to promote a sense of community.

Additionally, ethical considerations regarding back contamination are paramount. Article IX of the Space Treaty mandates controls to prevent both forward and backward contamination, necessitating strategies such as dry heat sterilization of spacecraft components and robotic assembly in sterile conditions. Sample return missions, classified as Category 5, require stringent precautions, following guidelines from NASA's Astrobiology program and the Chief Health and Medical Officer (Nicogossian et al., 2016, p. 90). Paul et al. (2022) noted that back contamination concerns from early Apollo missions were addressed through isolation protocols for astronauts and medical staff, as well as intensive containment measures for lunar samples at the JSC Lunar Receiving Laboratory. These strategies included high-level biological containment boxes and detailed searches for extraterrestrial pathogens, underscoring the importance of ethical considerations in ensuring the safety and integrity of both space and Earth.

Conclusion

NASA's structured three-phase plan for traveling from a lunar outpost to Mars and back emphasizes a comprehensive strategy for addressing the complexities of deep space exploration. By applying lessons learned from previous missions and fostering collaborations across various sectors, NASA is effectively tackling significant challenges related to transportation, operational efficiency, and astronaut health. The iterative nature of NASA's *Moon to Mars Architecture*, along with regular assessments and updates, allows for the continuous integration of technological advancements into their mission framework.

As humanity embarks on long-term lunar living and interplanetary exploration, understanding the health risks associated with space travel becomes increasingly important. Challenges such as radiation, cardiovascular issues, and muscle degeneration necessitate effective solutions to ensure the well-being of astronauts and future Martian residents. The rise of commercial space travel further underscores the need for robust space medicine practices to protect both professional astronauts and civilian tourists. Balancing security with ethical responsibilities is vital to ensure that our actions in space promote peace and sustainability. With a clear roadmap in place, NASA is well-prepared to achieve its ambitious goal of landing humans on Mars.

Limitations and Bias

There are considerable research gaps to this Mars trip analysis, and they are further exacerbated by limited funding and shifting political goals. The three key challenges addressed above are limited by the current inability to perform extensive research in deep space. Significant knowledge gaps exist in space medicine and human health, particularly regarding long-term effects with novel stimuli like lunar dust exposure. With fewer than 3,000 people having been exposed to spaceflight, the data available cannot yet be considered statistically robust. Increased investment in comprehensive studies is crucial to improving safety and the probability of mission success.

Continually, there may be ethical challenges surrounding medical policy formation and the exploration of other ecosystems. Much of the existing clinical knowledge about radiobiological effects comes from terrestrial exposure events, such as those experienced by atomic bomb survivors. Emerging evidence suggests that particle radiation is more harmful than electromagnetic radiation (Montesinos et al., 2021). While predicting the effects of long-term exposure to GCR and SPE is complex and necessitates further research, ethical and financial constraints rightfully impede swift progress in this area.

References

Basner, M., Dinges, D. F., Mollicone, D. J., Savelev, I., Ecker, A. J., Antonio, A. D., Jones, C. W., Hyder, E. C., Kan, K., & Morukov, B. V. (2014). Psychological and behavioral changes during confinement in a 520-day simulated interplanetary mission to mars. *PloS One*, *9*(3), e93298. https://doi.org/10.1371/journal.pone.0093298

Baiocco, G., Giraudo, M., Bocchini, L., Barbieri, S., Locantore, I., Brussolo, E., Giacosa, D., Meucci, L., Steffenino, S., Ballario, A., Barresi, B., Barresi, R., Benassai, M., Ravagnolo, L., Narici, L., Rizzo, A., Carrubba, E., Carubia, F., Neri, G., & Ottolenghi, A. (2020). The PERSEO experience: A water-filled garment prototype for personal radiation protection of astronauts successfully tested on board the International Space Station. *Aerotecnica, Missili e Spazio, 99*(2), 111–114. https://doi.org/10.1007/s42496-020-00048-0

Bourguignon, M. H., Gisone, P. A., Perez, M. R., Michelin, S., Dubner, D., Giorgio, M. D., & Carosella, E. D. (2005). Genetic and epigenetic features in radiation sensitivity: Part II: Implications for clinical practice and radiation protection. *European Journal of Nuclear Medicine and Molecular Imaging, 32*(3), 351–368. https://doi.org/10.1007/s00259-004-1731-6

Cheshier, L. (Host). (2024, December 20). Moon to Mars: Reviewing the architecture (No. 366) [Audio podcast episode]. In *Houston We Have a Podcast*. https://www.nasa.gov/podcasts/houston-we-have-a-podcast/moon-to-mars-reviewing-the-architecture/

Krichevsky, S. (2021). Artificial gravity for humans in space: Philosophical and methodological, historical, technological, and futurological aspects. *Philosophy and Cosmology: The Journal of the International Society of Philosophy and Cosmology (ISPC), 27*(27), 58–71. https://doi.org/10.29202/phil-cosm/27/4

NASA. (2015, October). Journey to Mars: Pioneering next steps in space. https://www.nasa.gov/wp-content/uploads/2017/11/journey-to-mars-next-steps-20151008_508.pdf?emrc=e76990

NASA. (2024, November 18). Moon to Mars architecture. https://science.nasa.gov/planetary-science/programs/mars-exploration/science-goals/

NASA. (2024, December 13). NASA's Moon to Mars architecture: Architecture definition document (Rev B). https://www.nasa.gov/wp-content/uploads/2024/12/esdmd-001-add-rev-b.pdf?emrc=194b15

National Aeronautics and Space Administration Authorization Act of 2010, S. 3729, 111th Cong. (2010). https://www.congress.gov/bill/111th-congress/senate-bill/3729

Nicogossian, A. E., Williams, R. S., Huntoon, C. L., Doarn, C. R., Polk, J. D., & Schneider, V. S. (2016). *Space physiology and medicine from evidence to practice*. Springer New York.

Nicogossian, A. (n.d.). Introduction to space medicine and human factors. SPST619. American Public University System. Lesson 1-8. Retrieved December 3, 2024, from https://myclassroom.apus.edu/d2l/le/sequenceLauncher/301704/View

McCoy, M. A., Liverman, C. T., & Kahn, J. (2014). *Health standards for long duration and exploration spaceflight: Ethics principles, responsibilities, and decision framework* (1st ed.). National Academies Press. https://doi.org/10.17226/18576

Miranda, S., Marchal, S., Cumps, L., Dierckx, J., Krüger, M., Grimm, D., Baatout, S., Tabury, K., & Baselet, B. (2023). A dusty road for astronauts. *Biomedicines, 11*(7), 1921. https://doi.org/10.3390/biomedicines11071921

Montesinos, C. A., Khalid, R., Cristea, O., Greenberger, J. S., Epperly, M. W., Lemon, J. A., Boreham, D. R., Popov, D., Gorthi, G., Ramkumar, N., & Jones, J. A. (2021). Space radiation protection countermeasures in microgravity and planetary exploration. *Life* (Basel, Switzerland), *11*(8), 829. https://doi.org/10.3390/life11080829

Paul, A.-L., Elardo, S. M., & Ferl, R. (2022). Plants grown in Apollo lunar regolith present stress-associated transcriptomes that inform prospects for lunar exploration. *Communications Biology, 5*(1), 382. https://doi.org/10.1038/s42003-022-03334-8

Pernigoni, L., Lafont, U., & Grande, A. M. (2024). Assessment of radiation shielding properties of self-healing polymers and nanocomposites for a space habitat case study under GCR and LEO radiation. *CEAS Space Journal, 16*(5), 525–533. https://doi.org/10.1007/s12567-023-00525-9

Rittweger, J., Albracht, K., Flück, M., Ruoss, S., Brocca, L., Longa, E., Moriggi, M., Seynnes, O., Di Giulio, I., Tenori, L., Vignoli, A., Capri, M., Gelfi, C., Luchinat, C., Franceschi, C., Bottinelli, R., Cerretelli, P., & Narici, M. (2018). Sarcolab pilot study into skeletal muscle's adaptation to long-term spaceflight. *NPJ Microgravity, 4*(1), 18. https://doi.org/10.1038/s41526-018-0052-1

Stough, C., et al. (2006). *Medical and health policy formulation for human exploration of the solar system: Part I: Ethical considerations in scarce resource allocation for health and medical systems development and operation.* George Mason University.

Tanaka, K., Nishimura, N., & Kawai, Y. (2017). Adaptation to microgravity, deconditioning, and countermeasures. *The Journal of Physiological Sciences: JPS, 67*(2), 271–281. https://doi.org/10.1007/s12576-016-0514-8

United States creates the U.S. Space Command and the U.S. Space Force to strengthen military capabilities in space. (2020). *The American Journal of International Law, 114*(2), 323–326. https://doi.org/10.1017/ajil.2020.13

Weisbin, C. R., Lavery, D. B., & Rodriguez, G. (1997). Robots in space: U.S. missions and technology requirements into the next century. *Autonomous Robots, 4*(2), 159–173. https://doi.org/10.1023/A:1008808730915

Why We "craft": The Significance of Narratives in Aerospace Healthcare and Beyond

Jeff Thompson, PhD
Edinburgh Napier University, School of Health and Social Care
Lipscomb University, College of Leadership and Public Service
Mediator.jeff@gmail.com

Abstract

This article advocates for the adoption of evidence-based, and, crucially, practical approaches to crafting personal narratives as a means of addressing the professional and personal challenges aerospace medical and healthcare professionals experience. Crafting personal narratives and utilizing them while supporting others can foster positive upward spirals that can both enhance professional competencies and be supportive of personal well-being. Instead of seeing narrative craft practices as a distraction, embracing this approach encourages a transformation working in these high-stress and other related work environments moving beyond merely surviving to genuinely thriving.

Keywords: Aerospace medicine, narrative psychology, narrative medicine, well-being, mental health, story-telling

Por qué "creamos": La importancia de las narrativas en la atención médica aeroespacial y más allá

Resumen

Este artículo aboga por la adopción de enfoques basados en la evidencia y, fundamentalmente, prácticos para la elaboración de narrativas personales, con el fin de abordar los desafíos profesionales y personales que enfrentan los profesionales de la medicina y la salud aeroespacial. Elaborar narrativas personales y utilizarlas al apoyar a otros puede fomentar espirales de progreso positivos que pueden mejorar las competencias profesionales y contribuir al bienestar personal. En lugar de considerar las prácticas de elaboración de narrativas como una distracción, adoptar este enfoque fomenta una transformación en estos entornos laborales de alto estrés y otros similares, que va más allá de la simple supervivencia para alcanzar un verdadero éxito.

doi: 10.18278/sesa.6.1.4

Palabras clave: Medicina aeroespacial, psicología narrativa, medicina narrativa, bienestar, salud mental, narración de historias

我们为何要“精心打造”：叙事在航空航天医疗保健及其他领域的重要性

摘要

本文倡导采用循证且至关重要的实用方法来构建个人叙事，以此应对航空航天医疗保健专业人员在职业和个人方面所面临的挑战。构建个人叙事并在支持他人的过程中加以运用，能促进积极的良性循环，既能提升专业能力，又能促进个人福祉。与其将叙事创作视为一种干扰，不如拥抱这种方法，它能帮助人们在高压及其他相关工作环境中实现转变——从生存转变为真正蓬勃发展。

关键词：航空航天医学，叙事心理学，叙事医学，福祉，心理健康，讲故事

Due to the nature of their work, aerospace medical and healthcare-related professionals are at risk suffering, when compared to the general public, increased mental health conditions, substance use disorders, suicide rates, and the pervasive effects of stigma—manifesting in reluctance to seek help and increased burnout. These concerns are not unique to aerospace healthcare professionals; they also affect other high-risk, high-pressure, help-providing professionals, including more broadly physicians (Andrew & Brenner, 2022) and nurses (National Academies of Sciences, Engineering, and Medicine, 2021), first responders (Jacobowitz et al., 2024), and astronauts (Arone et al., 2021).

This article advocates for the adoption of evidence-based, and, crucially, practical approaches to crafting personal narratives as a means of addressing the challenges that these aerospace medical and healthcare-related professionals are facing. Recent related studies demonstrate that by fostering positive upward spirals, such narrative practices can both enhance professional competencies and support their well-being (Launer & Wohlmann, 2023; Loy & Kowalski, 2024; Thompson & Launer, 2025; Thompson, 2023; Thompson & Jensen, 2023; Zaharias, 2018). This shift encourages a transformation in these high-stress aerospace work environments moving beyond merely surviving to genuinely thriving.

In the context of narrative medicine, and more broadly, narrative psychology and narrative health, crafting could appear synonymous with other story-development terms such as writing, telling, building, or creating. However, this article contends that crafting carries a distinct and purposeful significance, particularly in the aerospace healthcare profession, and its relevance extends well beyond this specific domain. As Dr. Rita Charon explains, the practice of narrative medicine represents "a commitment to understanding patients' lives, caring for the caregivers, and giving voice to the suffering" (Krisberg, 2017, para. 8ⁱ). More broadly, in literary terms, crafting has been described as "weaving evidence into a compelling narrative, writers make complex ideas accessible, create coherence and continuity, provide context and meaning, and make arguments more persuasive" (Mintz, 2024, para. 17).

In healthcare settings, crafting a narrative entails the intentional development of a personal story that honors the humanity and dignity of the individual, the others involved, and acknowledges the witnesses to the experience. This narrative crafting process can take multiple literary and creative forms, including written stories, poetry, visual art, and digital media (Charon, 2001; Charon et al., 2016). Participants are encouraged to engage in introspective reflection, write freely or in response to prompts, and participate in group discussions.

Far from being a distraction in clinical encounters, personal narratives are an essential component of aerospace healthcare practice and broader healthcare too—particularly when supporting individuals who are suffering. As the adapted wisdom of Sir William Osler reminds us, a good healthcare professional treats the disease, but a great one treats the person who is suffering (Centor, 2007).

Suffering, along with conflict, grief, anguish, and despair, are often embedded in the narratives emerging from aerospace medicine as well as healthcare, first responders, and others working in high-stress, care-oriented environments. Yet through purposeful crafting, these stories can also inspire hope, optimism, meaning, and purpose. They can evoke emotions such as gratitude, awe, and a sense of connectedness. When narratives are crafted with this degree of intention, the focus shifts away from rigidity, binary judgements, or over-analysis. Instead, as John Launer (2021, p.3) suggests, we adopt a stance of curiosity–moving from the question "What is really going on here?" to "How are people giving an account of their experience?" Importantly, this perspective applies both to our own stories and to the stories that we are exposed to.

Aerospace healthcare professionals are uniquely positioned to help others craft personal narratives that enable those they are supporting, the "authors" of their own experiences, to make sense of uncertainty, counter burnout and cynicism, derive meaning from ambiguity, and construct a more coherent understanding of their evolving life stories—all necessary for aerospace professionals, including but not limited to, astronauts and those working in analog environments. This process can also illuminate personal resilience and deepen a sense of connection with others.

Launer (2008) describes this practice as *Conversations Inviting Change*, a process that inspires hope in individuals and provides practical, supportive engagement, even in peer-to-peer interactions among colleagues.

For those suffering or struggling, the healthcare professional plays a vital and honored role as "co-author" or guide, helping to shape that part of the individual's life journey (Morgan, 2000). In this role, the professional facilitates the crafting of a re-storied narrative—one that is coherent, expresses self-efficacy, and affirms social connectedness.

It is important to emphasize that this work is not limited to healthcare professionals alone working in aerospace. Peer support personnel, for example, can provide similarly therapeutic-like forms of assistance. While not a substitute for professional care, peer support can play a valuable role in fostering well-being and resilience. One of its most significant benefits is its potential to encourage potential help-seeking–an approach that has gained increasing traction across several aerospace professions, including the aviation industry,[1] law enforcement (Drew & Martin, 2023), medicine (Tolins et al., 2023), and nursing (Pace et al., 2020).

When narrative crafting is implemented in small group settings as part of professional healthcare development, it can simultaneously enhance clinical or professional competence and promote personal well-being. Such settings cultivate mutual recognition of the humanity shared between caregiver and care recipient (Loy & Kowalsky, 2024). Participating in narrative practices within a supportive community of peers fosters a sense of belonging, which is particularly effective in mitigating burnout, moral injury, and professional cynicism. These environments generate upward spirals of both normative competence (the requisite professional skills) and narrative competence, all within a psychologically safe and collegial space–one that fosters esprit de corps.

Crafting and sharing narratives, and listening to the narratives of others reveals how the core practices of narrative medicine—attention, representation, and affiliation (Charon, 2005)—extend beyond literary texts and definitions, demonstrating their practical and profound effect on all of those involved in the story and those now bearing witness to it. Supporting the argument that "stories are science" (Thompson, 2025), honoring the experiences and perspectives of patients and healthcare professionals can improve overall well-being (Loy & Kowalsky, 2024).

While this article is a proponent of the notion that "stories are science" and crafting both professional and personal narratives can contribute to enhancing aerospace professional skills

1 Draft Aerospace Medical Association (AsMA) Resolution 2024-01 - *Integrating a Program of Mental Health and Wellness into Safety Management Systems (SMS) Across Aviation* - was recently passed by the AsMA Council and has since been put to a general membership vote.

and thriving in life, it is not without its limitations. Honing one's craft in narratives, or narrative competence, is never intended to replace normative competence, or the proficiency in the professionals' skills that are requisite for healthcare and aerospace professionals (Thompson & Launer, 2025). Further, when engaging in sharing narratives that involve others, such as those involving patients, aerospace healthcare professionals need to be mindful to not violate the rights and privacy of others. Additionally, crafting narratives need to be part of a diverse array of resilience and professional skill-building practices and strategies. Much like how there is not a single cause to suicide and the complexity of mental health conditions, these well-being and professional development efforts must also be diverse.

Crafting narratives illustrates that stories do not merely recount people's lives, they create them. As aerospace healthcare professionals, and fundamentally as human beings, we can continually refine our narrative capacities. This process enables us to flourish in our privileged roles, supporting others through the complex terrain of human experience—on the ground, in the sky, and beyond.

Conflict of Interest

The author has no conflict of interest to report related to the writing and submission of this article including receiving no funding.

References

Andrew, L.B., & Brenner, B. E. (2022, July 13). *Physician Suicide*. Medscape. https://emedicine.medscape.com/article/806779-overview?form=fpf

Arone, A., Ivaldi, T., Loganovsky, K., Palermo, S., Parra, E., Flamini, W., & Marazziti, D. (2021). The Burden of Space Exploration on the Mental Health of Astronauts: A Narrative Review. *Clinical neuropsychiatry, 18*(5), 237–246. https://doi.org/10.36131/cnfioritieditore20210502

Centor, R. M. (2007). To be a great physician, you must understand the whole story. *Medscape General Medicine, 9*(1), 59. https://pmc.ncbi.nlm.nih.gov/articles/PMC1924990/

Charon. R. (2001). Narrative medicine: A model for empathy, reflection, profession, and trust. *JAMA, 286*(15), 1897–1902. doi:10.1001/jama.286.15.1897

Charon, R. (2005). Narrative Medicine: Attention, Representation, Affiliation. *Narrative, 13*(3), 261–270. http://www.jstor.org/stable/20079651

Charon, R., DasGupta, S., Hermann, N., Irvine, C., Marcus, R. R., Colsn, E. R., Spencer, D., Spiegel, M. (2016). *The Principles and Practice of Narrative Medicine.* Oxford Academic. https://doi.org/10.1093/med/9780199360192.001.0001

Drew, J.M., & Martin, S. (2023). Mental health and wellness initiatives supporting United States law enforcement personnel: The current state-of-play. *Journal of Community Safety and Well-Being, 8*(Suppl 1), S12–S22. https://doi.org/10.35502/jcswb.298

Jacobowitz, R., Nitza, A., Tobin, K., & Harzard, J. (2024). *New York State First Responder Mental Health Needs Assessment*. The Benjamin Center at SUNY New Paltz. https://www.governor.ny.gov/sites/default/files/2025-02/First-Responder-MHNA-Final-Report.pdf

Krisberg, K. (2017, March 28). *Narrative medicine: Every patient has a story.* AAMC.org. https://www.aamc.org/news/narrative-medicine-every-patient-has-story

Launer, J. (2008). Conversations inviting change. *Postgraduate Medical Journal, 84*(987), 4–5. https://doi.org/10.1136/pgmj.2007.067009

Launer, J., & Wohlmann, A. (2023). Narrative medicine, narrative practice, and the creation of meaning. *Lancet, 401*(10371), 98–99. https://doi.org/10.1016/S0140-6736(23)00017-X

Loy, M., & Kowalsky, R. (2024). Narrative Medicine: The Power of Shared Stories to Enhance Inclusive Clinical Care, Clinician Well-Being, and Medical Education. *The Permanente journal, 28*(2), 93–101. https://doi.org/10.7812/TPP/23.116

Mintz, S. (2024, June 14). *The Power of a Story Arc in Scholarly Writing.* Inside Higher Education. https://www.insidehighered.com/opinion/blogs/higher-ed-gamma/2024/06/14/how-scholarly-writing-can-transform-research-narratives

Morgan, A. (2000). *What is Narrative Therapy?* Dulwich Centre Publications.

National Academies of Sciences, Engineering, and Medicine. (2021). *The Future of Nursing 2020–2030: Charting a Path to Achieve Health Equity.* Washington, D.C.: The National Academies Press. https://doi.org/10.17226/25982.

Pace, E.M., Kesterson, C., Garcia, K., Denious, J., Finnell, D.S., & Bayless, S.D. (2020). Experiences and outcomes of nurses referred to a peer health assistance program: Recommendations for nursing management. *Journal of Nursing Management, 28*(1), 35–42. https://doi.org/10.1111/jonm.12874

Thompson, J. (2025). *Narrative psychology: Examining the science of awe stories in fostering resilience and wellbeing.* Thesis (PhD Doctorate). Edinburgh Napier, University, School of Health and Social Care. Edinburgh, Scotland.

Thompson, J., & Launer, J. (2025). Illuminating humanity: Narrative medicine practices for crisis negotiators. *Journal of Police and Criminal Psychology.* https://doi.org/10.1007/s11896-025-09756-4

Thompson, J. (2025). *Narrative psychology: Examining the science of awe stories in fostering resilience and wellbeing.* Thesis (PhD Doctorate). Edinburgh Napier, University, School of Health and Social Care. Edinburgh, Scotland.

Tolins, M. L., Rana, J. S., Lippert, S., LeMaster, C., Kimura, Y. F., & Sax, D. R. (2023). Implementation and effectiveness of a physician-focused peer support program. *PloS one, 18*(11), e0292917. https://doi.org/10.1371/journal.pone.0292917

Zaharias G. (2018). What is narrative-based medicine? Narrative-based medicine 1. *Canadian family physician Medecin de famille canadien, 64*(3), 176–180.

Evolution of the Human Space Program: The Human Element

Charles R. Doarn, MBA, FATA, FAsMA

Armstrong Institute for Space, Technology and Research, Office of Research and the Department of Environmental and Public Health Sciences, College of Medicine, University of Cincinnati, Cincinnati, Ohio, USA

Corresponding Author:
Charles R. Doarn, MBA, FATA, FAsMA
Director, Armstrong Institute for Space, Technology, and Research (ASTRO)
Office of Research
Emeritus Professor
Department of Environmental and Public Health Sciences
College of Medicine
University of Cincinnati
Cincinnati, OH 45267
Editor-in-Chief, *Telemedicine and e-Health Journal*
Cell: (513) 403-9604
E-mail: charles.doarn@uc.edu

Abstract

For millennia, humans have yearned to explore the unknown. Looking into the night sky, the thousands of generations before us, wondered what was out there. It was not until the 19th century that one could be above the ground in a balloon. The Wrights conquered flight in the early 20th century and Goddard developed rocketry in America in the 1920s—but it was a World War that brought rapid change. Operation Paperclip brought German scientists to the United States beginning in 1945 and they helped usher in the Space Age. The advent of human space flight began in earnest in 1958 and it eventually led to humans on the Moon, orbiting space stations, reusable spacecraft and commercial companies that ferry astronauts and supplies to the International Space Station. Over the past seven decades, we have evolved our space program from one of governmental priorities and investment to one of a burgeoning commercial program that now sends tourist above the Kármán line and into space.

Keywords: history, spaceflight, politics, exploration, aerospace, aviation

doi: 10.18278/sesa.6.1.5

Evolución del programa espacial humano: el elemento humano

RESUMEN

Durante milenios, la humanidad ha anhelado explorar lo desconocido. Al observar el cielo nocturno, las miles de generaciones anteriores se preguntaban qué había allí afuera. No fue hasta el siglo XIX que se pudo volar en globo. Los hermanos Wright conquistaron el vuelo a principios del siglo XX y Goddard desarrolló la cohetería en Estados Unidos en la década de 1920; sin embargo, fue una Guerra Mundial la que trajo consigo un cambio rápido. La Operación Paperclip trajo científicos alemanes a Estados Unidos a partir de 1945, quienes contribuyeron al inicio de la Era Espacial. El advenimiento de los vuelos espaciales tripulados comenzó en serio en 1958 y finalmente condujo a la llegada de humanos a la Luna, estaciones espaciales en órbita, naves espaciales reutilizables y compañías comerciales que transportan astronautas y suministros a la Estación Espacial Internacional. Durante las últimas siete décadas, nuestro programa espacial ha evolucionado, pasando de ser una prioridad gubernamental y de inversión a un floreciente programa comercial que ahora envía turistas por encima de la línea de Kármán hacia el espacio.

Palabras clave: historia, vuelos espaciales, política, exploración, aeroespacial, aviación

人类航天计划的演变：人的因素

摘要

千百年来，人类一直在探索未知领域。仰望夜空，我们之前的数千代人都在思考宇宙的奥秘。直到19世纪，人们才得以乘坐热气球升空。20世纪初，莱特兄弟实现了飞行，戈达德在20世纪20年代在美国发展了火箭技术——但真正带来快速变革的是一场世界大战。"回形针行动"于1945年开始，德国科学家来到美国，他们助力开启了太空时代。载人航天真正开始于1958年，最终实现了人类登月、空间站绕地运行、可重复使用航天器以及为国际空间站运送宇航员和物资的商业公司。过去七十年，我们的太空计划从政府的重点投资项目演变成为蓬勃发展的商业项目，后者如今已能将游客送上卡门线以上的太空。

关键词：历史，太空飞行，政治，探索，航空航天，航空

Introduction

From the beginning of human existence, our ancestors looked to the night sky and wonder what was there, and perhaps if they could ever go there. While it took millennia to develop technologies that improved human existence, it was not until we, as humans, harnessed energy and natural resources to create ways of getting off the ground.

While there is a rich history and a plethora of material on the history of the human spaceflight program in the literature, this short abridgement provides a summary of our desire to get off the ground and into space. It also provides a human element perspective of where we are going as a species.

Pre-World War II

From antiquity to the beginning of the modern era, a transformation occurred with each passing century, whereby new technologies, like gunpowder, and science fiction writers, dreamed of new worlds. In the 18th century, the Montgoflier Brothers built and flew hot air balloons. Étienne Montgolfier was the first human to actually lift off the earth. In the mid-19th century, British researchers James Glaisher and Henry Coxwell flew a balloon to 37,000 ft without supplement oxygen.[1] They both suffered significant challenges in physiological response to the change in pressure and temperature; laying the foundation for future aviation questions.[2]

At the beginning of the 20th century, the Wright Brothers, building on earlier work by Otto Lillenthal and others, developed and flew a powered biwing airplane, incorporating pitch, yaw, and role on the sand dunes near Kitty Hawk, North Carolina (See **Fig. 1**). This ushered in a new era where humans could go aloft in aeroplanes. Initially thought to be a skeptical adventure, the aeroplane became invaluable tool of war. A decade later, Robert Goddard was experimenting with liquid-fueled rockets. By the time World War II began in the late 1930s, the boundaries, capabilities and utility of balloons, airplanes, and rockets were expanding.

Fig. 1. Wright flier at Kitty Hawk 1903

Since human beings were in both balloons and airplanes, medical personnel began to develop a general understanding of how human physiology was impacted above 10,000 ft and below 35,000 ft. This new discipline, aviation medicine, began in 1918 with the United States Army School of Aviation Medicine, with a focus on the human in the system.

Beyond the Biosphere

In 1915, the National Advisory Committee on Aeronautics (NACA) was established.[3] This organization was the precursor to the National Aeronautics and Space Administration (NASA) focused on aviation and when NASA was formed, it absorbed thousands of NACA employees, assets, labs and buildings.

Toward the end of World War II, allied forces descended upon Peenemünde, in the far northern region of Germany. This was where the Nazi regime perfected the V2 rocket. The United States (U.S.) Army and the Red Army (USSR) were focused on capturing both technology and the scientists and engineers.[4] In 1945, the U.S. established a secret program, named Operation Paperclip, with the sole purpose of moving material (unused rockets) and more importantly over 1,600 captured German scientists, engineers, and technicians to the U.S. A key leader of this group, Dr. Wernher von Braun, was a vital German expert.[5] He eventually designed the Saturn V rocket, which would take Americans to the moon a mere 14 years later in 1969.

With rocket capabilities, albeit, in early development, the implication of sending humans into space was no longer an impossibility. There was limited knowledge on whether or how humans could survive in this austere environment.

In the 1950s, research was conducted on animals using intercontinental ballistic missiles. When the USSR launched Sputnik on October 4, 1957, America was caught off guard, prompting President Dwight Eisenhower to react. In 1958, NASA and the Defense Advanced Research Project Agency (DARPA) were established. NASA established the Space Task Group (STG) at Langley Research Center with a focus on the early Mercury Program (See **Fig. 2**), including vehicle design and crew selection.

While the U.S. and USSR had captured and patriated the German cadre of expertise, the issue of "could humans actually survive?" was front and center. In the U.S. all of the expertise in bioastronautics was within the U.S. Air Force (USAF), which itself was less than 10 years old, having been created out of the U.S. Army Air Corp in 1949. **Table 1** lists a number of questions experts had at the beginning of human space flight.

Fig. 2. NASA's Space Task Group 1959

Table 1. A partial list of unanswered question (ca 1960)

Question No	Question
1	What happens to the body in space?
2	Can man survive with all the stresses he will experience?
3	How to live and work in space?
4	How to monitor the human in this environment?
5	How to eat?
6	How to eliminate wastes?
7	What will acceleration do to the body?
8	What about radiation?
9	How to provide healthcare in space?

The launch of Sputnik and President Kennedy's directive to land a man on the moon and return him safely to the Earth, laid the foundation of the Space War, which of course was a significant element of the Cold War.

Prior to the launch of Yuri Gagarin on Vostok 1 in April 1961 and Alan Shephard on the Mercury-Redstone 3 in May 1961, the knowledge base of human capabilities in space was limited. Primate and other mammalian studies in the 1950s as well as the Manhigh Studies in the late 1950s provided a growing knowledge base of the physiological impacts. With each successive flight of astronauts and cosmonauts, it became apparent that while spaceflight affected the individual's physiology, they could survive and function to meet mission objectives.

There were several challenges as NASA began. Aside from political machinations, engineers and physicians were often at odds. In the early design phase of the Mercury capsule, the engineers placed a narrow window out of reach of the astronaut's view. The flight surgeons at the time argued that a window was necessary, positing that the astronaut could see the earth below. The engineers countered with doubt that they would be able to see anything. The flight surgeons, with concurrence from the astronauts, won out, and all spacecraft have windows that are accessible by the astronauts.

Exploration Paradigm

Once the Mercury program was completed, the Gemini and Apollo programs began to accelerate in development, including design, training and mission objectives. In the early 1960s, the USAF had developed the Manned Orbiting Laboratory (MOL), which was competing with NASA and of course access to funding, which was being directed to the conflict in Vietnam. The MOL astronauts transitioned to NASA in 1969 with several eventually flying in the Shuttle Program.

The Russians continued to pursue short missions with the Vostok, Voshkod, and Souyz including the first woman, Valentina Tereshkov. The Russians began to develop and maintain successive space stations in orbit (Salyut, DOS, and Mir) for longer duration missions with, where crews rotated after months on orbit. On the U.S. side, the longest missions, initially were on the Skylab in the early 1970s.

Each successive U.S. program built on the previous from an engineering perspective. The Mercury and Gemini programs provided new knowledge on crew health, but research was limited due to the limited workspace in the spacecraft. The objective of the human spaceflight program in the 1960s was to get a man on the moon.

The buildup of the Mercury and Gemini program entailed many new technologies, capabilities, and challenges. This included crew selection, crew observation using early telemedicine capabilities, tracking of the spacecraft, communications and recovery. **Fig. 3** illustrates the tracking stations around the world where medical observes were ensconced.

While there were early discussions between President Kennedy and Premier Nikita Kkrushchev about a joint mission to the moon, there was limited interests and each nation pursued their own paths.

Through many challenges, including the Apollo 1 (AS-204) fire, which claimed the lives of three astronauts (Roger Chaffee, Gus Grisson, and Ed White), NASA landed on the moon

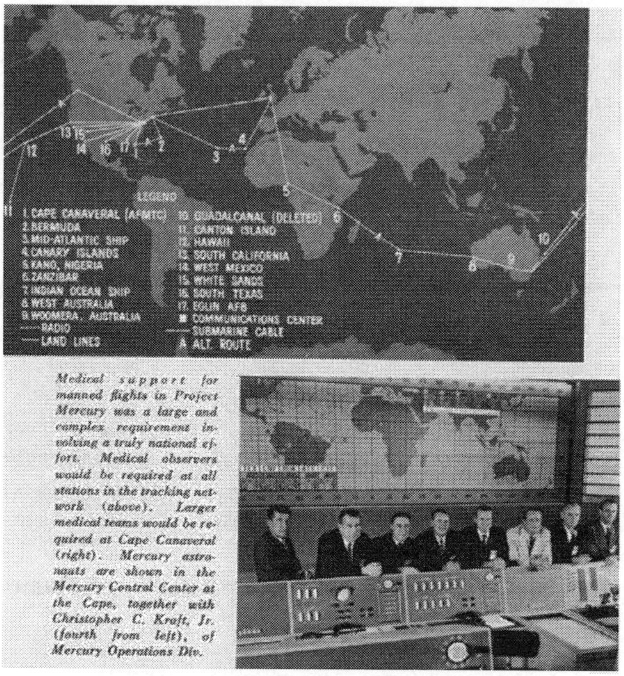

Fig. 3. Mercury medical support locations.

Fig. 4. Space Shuttle and crawler (courtesy of NASA).

Fig. 5. The International Space Station.

on July 20, 1969, fulfilling Kennedy's challenge. In the years that followed, additional missions landed on the moon and several brought back samples.

In NASA's formative years, the concepts of orbiting space station and reusable winged spacecraft were conceived. By the early 1980s, NASA had built and launched the Space Shuttle Columbia (STS-1) (See **Fig. 4**). From the human perspective, crew selection and medical care systems were developed without the lead of the USAF. In 1977, NASA established its own standards for crew selection and a medical policy board to review issues related to crew health.[6]

Permanent Presence

From the first Shuttle launch in 1981 to the last Shuttle launch in 2011, NASA flew 135 missions. The Challenger and Columbia were lost in horrific accidents. The Shuttle docked with the Mir Space Station in successive missions in the 1990s, built the International Space Station (ISS), and delivered or repaired various satellites, including the Hubble. (See **Fig. 5**)

For more than 25 years, humans from around the world have continuously lived and worked on the ISS conducting valuable research and increasing our knowledge of both human physiology and the natural sciences of space and of the Earth. In order to support the ISS, the lessons dating to the beginning of human spaceflight in the 1960s served as a foundation. Based on the U.S. and Russia efforts in early flight programs, the international community developed a framework that embodies collaboration for the common good.[7-11] This framework and knowledge has set the stage for new programs like Artemis, Gateway and eventual human exploration of Mars and other celestial bodies. The ISS has provided an opportunity to understand human challenges for durations that now measure more than a year. This will be of value in developing human presence on transit missions to Mars and surface exploration.

There are many unknown challenges for human mission to Mars, including radiation exposure, impact on human physiology, lack of real-time communications, resources (food, water, etc.), limited gravity, temperature extremes, operational tempo, and of course human psychology, tempered with ethics and isolation.

Conflict Management

As mentioned earlier, there is perceived conflict between various disciplines in support of human spaceflight. This paradigm goes back to the very beginning of NASA in 1958. The STG, which was relocated Houston, TX as the Manned Spacecraft Center believed it needed little impact from NASA Headquarters in Washington, D.C. The physicians and engineers were often in conflict as they viewed the human in the loop differently.[12] In 2016, NASA developed a book that reviewed these challenges.[13]

Engineers...	Life Scientists...
...design their systems	...reverse engineer Nature to understand the system
...use quality controlled components	...study diverse individuals with diverse components
...use established frameworks to employ physical laws	...discover concepts and qualitative relationships before developing quantitative understanding
...design PID controllers	...discover biased walks in chemotaxis

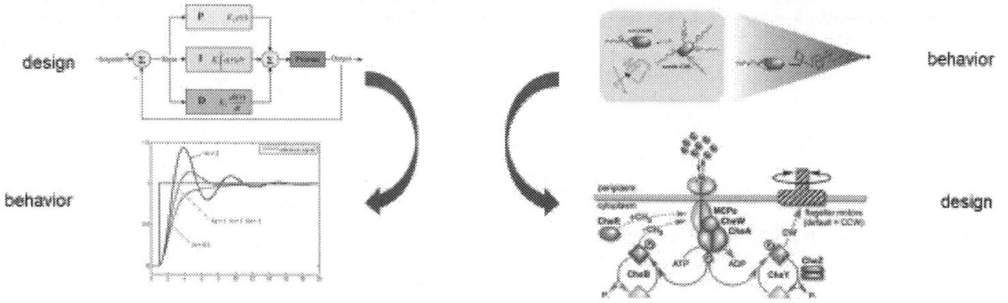

Fig. 6. Differences between Life Sciences and Engineering.

Fig. 6 illustrates the differences between engineers and life scientists. Simply put, engineers can test things to destruction and in medicine, this is not the approach. Over the past seven decades, much of this dichotomy has been ameliorated in part due to effective communication, the development of health and medical technical authority in the aftermath of the Columbia accident, and physician astronauts.

In international programs, such as the Apollo-Soyuz Test Project in 1975 and the ISS Program, highly effective frameworks were established to support design and human health.[8]

Commercial Paradigm

For most of the history of human spaceflight, the purview has been government-funded and operated systems. In 1985, President Ronald Reagan signed the Commercial Space Launch Act of 1984 to support commercial spaceflight, which was amended in 2004 and signed into law by President George W. Bush. These laws provided the foundation for investment and launch services for commercial companies to serve NASA and paying customers to travel to space. Governed by the Federal Aviation Administration, commercial companies have rapidly expanded capabilities in support of human spaceflight.

Companies like SpaceX, Blue Origins, and Virgin Galactic, each with unique platforms have sent paying customers into space. These companies have built successful models based on the experience and lessons learned over the past seven decades. While the future is unwritten, it will be a collective effort by both governments and commercial entities that will expand the horizon for human spaceflight. Perhaps one day, in the not too distant future, the number of individuals flying to Earth-orbiting platforms for a long weekend, may be similar to taking a commercial aircraft to a destination on the ground.

The Next Giant Leap

Human exploration of the Moon and Mars are now a central focus of the exploration narrative by NASA, other nations, and commercial companies. Challenges and unanswered questions remain, especially with regards to Mars.[14] The foundations of human spaceflight are built on the experiences and lessons learned by those men and women who have supported programs or have flown in space.

Once humans escaped the surly bonds of earth and ventured in space, the concept of traveling to distant planets was no longer science fiction. There is no doubt the challenges of a human mission to Mars would be immensely challenging. But as President Kennedy espoused so long ago, "*...not because they are easy, but because the hard ....*" This may very well become the greatest challenge of the 21st century. Much of the technology to support the transportation of such and endeavor exists. It is the human element and support of the crew that will be a far greater challenge.

Notes

1 Doherty MJ. James Glaisher's 1862 account of balloon sickness: altitude, decompression injury, and hypoxemia. *Neurology*. 2003;60(6): 1016–8. doi: 10.1212/01.wnl.0000048663.44594.a2.

2 West JB. Paralysis and blindness during a balloon ascent to high altitude. High Alt Med Bio 2004; 5: 453–55.

3 Orders of Magnitude: The History of the NACA and NASA, 1915–1990. https://www.nasa.gov/wp-content/uploads/2023/03/sp-4406.pdf. [Last accessed: October 1, 2025]

4 Gimbel J. Project Paperclip: German Scientists, American Policy, and the Cold War. *Diplom Hist*. 1990;14(3): 343–366, https://doi.org/10.1111/j.1467-7709.1990.tb00095.x

5 Emme, E. M. History of Rocketry and Space Travel by Wernher von Braun, Frederick I. Ordway III. *Technology and Culture*. 1968; 9(2): 250–1.

6 Doarn, C. R. Medical policy development for human spaceflight at NASA: An evolution. *Aviat Space Environ Med*. 2011 Nov; 82(11): 1073–7. doi: 10.3357/asem.3103.2011.

7 Grigoriev, A.I., Williams, R.S., Comtois, J-M, et al. Space Medicine Policy Development for the International Space Station. *Acta Astronautica*. 2009; 65(5-6): 603–12.

8 Doarn, C.R., Polk, J. D., Grigoriev, A., et al. A Framework for Multinational Medical Support for the International Space Station: A Model for Exploration. *Aerosp Med Hum Perform*. 2021 Feb 1; 92(2): 129–134.

9 Bogomolov, V. V., Castrucci, F., Comtois, J. M., et al. International Space Station med-

ical standards and certification for space flight participants. *Aviat Space Environ Med.* 2007 Dec; 78(12): 1162–9. doi: 10.3357/asem.2175.2007.

10 Morgun, V. V., Voronin, L. I., Kaspranskiy, R. R., et al. Development of U.S.-Russian medical support procedures for long-duration spaceflight: the NASA-Mir experience. *Aviat Space Environ Med.* 2002 Feb; 73(2): 147–55.

11 Duncan, J. M., Bogomolov, V. V., Castrucci, F. et al. Organization and Management of the International Space Station (ISS) multilateral medical operations. *Acta Astronautica.* 2008; 63(7-10): 1137–147.

12 Doarn, C.R. Evolution of Space Medicine at NASA. *Aerosp Med Hum Perform.* 2024 Oct; 95(10): 797–805. doi: 10.3357/AMHP.6472.2024.

13 Doarn, C. R., Travis, T. W., Currie-Gregg, N.K., et al. Engineering, Life Sciences, and Health/Medicine Synergy in Aerospace Human Systems Integration: The Rosetta Stone Project: An Executive Summary. *New Space.* 2019; 7(2): 110–13.

14 Doarn, C. R., Orlov, O., Kussmaul, A., et al. Challenges in Providing Medical Support on an International Human Expedition to Mars: An Overview. *Acta Astronautica.* 2025; 235: 691–695.

ASTP: The Beginning of 50 Years of East-West Collaboration in Outer Space

Arnauld Nicogossian, MD, MS, FAsMA, FACPM, FACP[1, 2]

Charles R. Doarn, MBA, FATA, FAsMA[3,4]

[1] Schar School of Policy and Government, George Mason University, Arlington, Virginia, USA;
[2] American Public University, Charlestown, West Virginia, USA;
[3] Armstrong Institute for Space, Technology and Research, Office of Research and
[4] Department of Environmental and Public Health Sciences, College of Medicine, University of Cincinnati, Cincinnati, Ohio, USA

Corresponding Author:
Arnauld Nicogossian, MD, FAsMA, FACPM, FACP
Affiliate Faculty Schar School of Policy and Government
George Mason University
Arlington, Virginia
Part Time Faculty APUS and ASTP NASA Crew Surgeon
E-mail: arnauld.nicogossian@mycampus.apus.edu

Limitations and Bias: Availability of peer reviewed publications is limited. The medical results identify each astronaut by name or crew position and in conformance with the NASA guidelines for release of public information following mission anomaly or crew injury. Most of the ASTP mission and experiments results were published on government websites.

Abstract

At the end of World War II, the allied alliance was replaced by the East and West "Cold War" marked by the United States (U.S.) and the Union of Soviet Socialist Republics (USSR) rivalry on Earth and in space. The launch of the Sputnik 1, followed by Cosmonaut Yuri Gagarin spaceflight, shook the American political and technological confidence and launched the "Space Race." The space rivalry lasted 25 years, until the start of the Détente [1967–1979]. One of the first highly publicized peaceful collaboration was a joint ASTP space mission and the launch of the U.S./USSR Joint Working Group on Space Biology and Medicine. Apollo and Soyuz space crafts were launched on July 15, 1975, and docked on July 17, 1975. The space crafts remained docked for two days with multiple successful crew transfers using the *Docking Module*. The crew

 doi: 10.18278/sesa.6.1.6

completed 28 experiments before returning to Earth but was exposed to a highly toxic nitrogen oxide [N_2O_4] compound used as an oxidizer of the Apollo command module thrusters. The crew developed a chemical pneumonitis requiring hospitalization and treatment at the Tripler Army Medical Center in Honolulu, Hawaii before returning to the NASA Lyndon B. Johnson Space Center in Houston, Texas. ASTP was a demonstration of the rendezvous and docking systems and space rescue. An important outcome of ASTP was to serve as a template for future more ambitious international collaboration in space. This collaboration survived many political East-West tensions.

Keywords: history, Apollo-Soyuz, international collaboration, space docking, space rescue

ASTP: El comienzo de 50 años de colaboración Este-Oeste en el espacio exterior

Resumen

Al final de la Segunda Guerra Mundial, la alianza aliada fue reemplazada por la "Guerra Fría" entre Oriente y Occidente, marcada por la rivalidad entre Estados Unidos (EE. UU.) y la Unión de Repúblicas Socialistas Soviéticas (URSS) en la Tierra y en el espacio. El lanzamiento del Sputnik 1, seguido del vuelo espacial del cosmonauta Yuri Gagarin, sacudió la confianza política y tecnológica estadounidense e inició la "Carrera Espacial". La rivalidad espacial duró 25 años, hasta el inicio de la Détente [1967-1979]. Una de las primeras colaboraciones pacíficas altamente publicitadas fue una misión espacial conjunta ASTP y el lanzamiento del Grupo de Trabajo Conjunto EE. UU./URSS sobre Biología y Medicina Espacial. Las naves espaciales Apolo y Soyuz fueron lanzadas el 15 de julio de 1975 y acopladas el 17 de julio de 1975. Las naves espaciales permanecieron acopladas durante dos días con múltiples transferencias exitosas de tripulación utilizando el Módulo de Acoplamiento. La tripulación completó 28 experimentos antes de regresar a la Tierra, pero estuvo expuesta a un compuesto altamente tóxico de óxido de nitrógeno [N_2O_4], utilizado como oxidante de los propulsores del módulo de mando del Apolo. La tripulación desarrolló una neumonitis química que requirió hospitalización y tratamiento en el Centro Médico Tripler del Ejército en Honolulu, Hawái, antes de regresar al Centro Espacial Lyndon B. Johnson de la NASA en Houston, Texas. El ASTP fue una demostración de los sistemas de encuentro y acoplamiento, así como del rescate espacial. Un resultado importante del ASTP fue servir de modelo para futuras colaboraciones internacionales más ambiciosas en el espacio. Esta colaboración sobrevivió a numerosas tensiones políticas entre Oriente y Occidente.

Palabras clave: Historia, Apolo-Soyuz, colaboración internacional, acoplamiento espacial, rescate espacial

阿波罗-联盟测试计划：东西方在太空领域50年合作的开端

摘要

二战结束后，同盟国被东西方“冷战”所取代，其标志是美国和苏联在地球和太空的竞争。苏联发射人造卫星1号，随后宇航员尤里·加加林进入太空，动摇了美国的政治自信和技术自信，并开启了“太空竞赛”。这场太空竞赛持续了25年，直到缓和时期（1967-1979年）的开始。最早备受瞩目的和平合作之一是阿波罗-联盟测试计划(ASTP)任务以及美苏空间生物学和医学联合工作组的成立。阿波罗号和联盟号飞船于1975年7月15日发射升空，并于1975年7月17日完成对接。飞船对接持续了两天，期间利用对接舱成功进行了多次宇航员转移。在返回地球前，宇航员完成了28项实验，但却暴露于一种剧毒的氮氧(N_2O_4)化合物中，这种化合物被用作阿波罗指令舱推进器的氧化剂。宇航员患上了化学性肺炎，需要在夏威夷檀香山的特里普勒陆军医疗中心接受住院治疗，之后返回位于德克萨斯州休斯顿的美国宇航局林登·约翰逊航天中心。ASTP是对交会对接系统和太空救援的演示。ASTP的一个重要成果是为未来更具雄心的国际太空合作树立了典范。这项合作经受住了东西方之间诸多政治紧张局势的考验。

关键词：历史，阿波罗-联盟测试计划，国际合作，太空对接，太空救援

Background

Prior to the Apollo Soyuz Test Project (ASTP), NASA robotic missions did include international participation, which was not the case for crewed missions. The exception was the flight of the *Biostack* device, designed to study the biological effects of high energy cosmic radiation (HZE/GCR) were flown on Apollo 16 and 17 missions (Facius et al. 1979; Horneck 1993).

In 1945, with the end of World War II, the East-West alliance was replaced by the "Cold War," marked by the United States (U.S.) and the Union of Soviet Socialist Republics (USSR) rivalry on Earth and in Space and by regional conflicts (Schlesinger 1967; Westad 2017). The launch of the Sputnik 1, followed by Cosmonaut Yuri Gagarin space flight, challenged the American political and technological leadership on a global scale launching the "Space Race" (Devezas et al. 2012; Orr 2004). The détente timelines and ASTP agreement are shown in **Table 1.**

Table 1. The Détente Timelines 1969–1979

Policy Level	**Space Agencies (ASTP)**
Nixon/Brezhnev U.S.-USSR summit meetings held in Moscow on May 24, 1972, and Washington June 25, 1973	NASA visit to USSR 1971
Agreements on nuclear arms control, space and health collaboration	U.S./USSR JWG on Space Biology and Medicine 1971
President Richard Nixon and Premier Alexei Kosygin also signed an agreement calling for an Apollo-Soyuz mission for July 1975	NASA and USSR Academia of Sciences with Chairs: Glynn Lunney (NASA) and Konstantin Bushuyev (Director, Energia RKK)
The cooperation renewed in 1989 (Russia leaving Afghanistan)	Three years of training 1972–1975
	• USSR: Star City, Baikonur, TsNIIMASH/TSUP
	• U.S.: JSC, KSC, and aerospace companies

The start of the Détente [1967–1979] ended 25 years of space competition and heralded the beginning of a new era of nuclear arms control and collaboration in space symbolized by the "*hand shake in space.*" The ASTP was one of the center pieces of this collaboration. Another but less publicized event was the launch of the U.S./USSR Joint Working Group on Space Biology and Medicine (Doarn et al. 2010).

The primary focus of the ASTP was to showcase the U.S. and the USSR in the new era of détente and demonstrate the ability of on orbit render help and rescue of astronauts and cosmonauts (Ross-Nazzal 2010). The mission was also a first demonstration of the peaceful use of space consistent with the Outer Space Treaty (OST) principles (Hall 1969). Supplementary objectives consisted of conducting engineering, physical, and biomedical experiments (El-Baz and Warner 1977; Fitts et al. 2011).

Method

Google Scholar, Web of Science, PubMed were used as primary search engines using terms [but not limited to] such as "international collaboration in space; U.S. and USSR collaboration in space; ASTP mission requirements; bends prevention; space medicine; NASA-USSR joint space experiments; ASTP docking module; crew safety; nitrogen tetroxide inhalation; and ASTP crew post mission medical care." Preference was given to peer reviewed articles in English language published since 1912.

Results

Abstracts from 73 publications of interest were reviewed. A total of 28 publications met the inclusion criteria. Seven publications are either conference papers, websites, books or NASA Technical Reports (El-Baz and Warner 1977; Facius et al. 1977; Page and Page 1977; Kay 1998; Sandler and Grigoriev 1990; Westad 2017; Hanhimäk 2012) were judged influential and included in the reference list. The remainder are peer-reviewed publication considered as both influential and robust evidence.

Discussion

During the "Cold War," the U.S. President John F. Kennedy's offer to the USSR for a joint exploration of the Moon (Kay 1998) partnership was not realized. During the "Cold War," President Kennedy's offer to USSR for a joint exploration of the Moon (Kay 1998) partnership was not realized. Both nations pursued their space activities independently, with the U.S. landing astronauts on the Moon and operating three Skylab missions in low Earth orbit, while the Russians operated orbital Salyut stations.

It was not until the early 1970 that President Nixon's desire to reduce the threat of a nuclear war created the opportunity for scientific collaboration, with the USSR including a joint space mission (Zelizer 2009).

The ASTP envisioned docked Apollo and Soyuz spacecrafts with three astronauts Thomas Stafford, Donald "Deke" Slayton and Vance Brand to meet Russian cosmonauts Aleksey Leonov and Valeriy Kubasov as a center piece of this "détente" policy.

Mission Docking Requirements

- All transfers between spacecraft will be in "shirtsleeves." Crewmen will sleep in their own spacecraft.
- Docking module systems will be operated by an Apollo crewman, and two men will be in the docking module during a transfer operation.
- Spacecraft hatches. will be operated by the crew of that spacecraft,
- No provision for contingency.
- Each Apollo crewman and each Soyuz crewman will visit the other spacecraft at least once during the two days of docked operations.
- During transfers at least one host crew member in each space craft and respective spacecraft hatch to the docking module is close.

1. ***Pre-flight Crew Health***

    a. Donald "Deke" Slayton, one of the seven original Mercury astronauts was removed from the flight status in 1962 after being diagnosed with a benign idiopathic intermittent atrial fibrillation.

    b. After exhaustive medical evaluations by the then leading cardiologists, Mr. Slayton was reinstated to flight status (aviation).

    c. Mr. Slayton was assigned to the last Apollo mission (ASTP) in 1970.

    d. During the pre-flight and through the ASTP mission Mr. Slayton remained free of atrial fibrillation (Nicogossian 1977).

2. ***Transfer and docking module***

    a. The docking on orbit and crew transfers was required to be in a shirt sleeve environment. The Apollo command module was designed to operate at 1/3 atmosphere and 100% O_2, while the Soyuz capsule was pressurized to one atmosphere with 21% O_2 and 79% N_2. The pressure and gas composition of the two spacecrafts were incompatible with his requirement and presented a danger for decompression sickness.

    b. To meet the mission requirements, NASA designed a docking module, built by the Rockwell International Corporation to allow the docking of the spacecrafts using the standard Apollo probe and drogue docking system and the Russian androgynous docking mechanism designated as the APAS-75 (Lewis and Donahoe 2023).

    c. The docking module also acted as an airlock to prevent bends and ensure a safe crew transfer.

    d. Multiple altitude chamber studies at the Institute for Biomedical Problems, in Moscow, USSR, United States Air Force Brooks Medical Laboratory in San Antonio, Texas, and the NASA Lyndon B. Johnson Space Center, Houston, Texas with a pool of over 41 subjects were used to validate a bends free protocol. The final procedure required the Soyuz spacecraft to drop its pressure to 10.2 psia [527.5 mmHg] during the docking and transfer allowing crews to equalize the pressure and gas composition during transfers (Nicogossian and

Campbell 2023).

e. During the 44 hours of the docked time four transfers were completed with the participation of all five crewmembers. The docking module protocol safety was validated without either cosmonauts or astronauts suffering from decompression sickness (Weathersby et al. 1984).

f. No medical issues were encountered during the flight portion of ASTP.

3. **Post-flight**

What started as a celebrated and successful space mission ended up as a medical emergency. During the reentry into the Earth's atmosphere the Apollo Command Module (CM) reaction control system (RCS) thrusters were not turned off. Nitrogen oxide [N_2O_4] entered the CM exposing the crew for at least four minutes to the toxic compound. The mission commander acted rapidly and after the Pacific Ocean splashdown administered oxygen to himself and the other two crewmembers. A short ceremony was held in the hangar deck of the recovery USN LPH 1. Toward the end of a telephone call with President Gerald Ford, the ASTP commander informed the crew surgeon of the toxic "smoke" exposure. The crew was promptly showered, administered steroids and moved to the LPH 11 dispensary for further care and observations.

All post-flight planned medical evaluations and on deck ceremonies were cancelled. Twelve hours later the crew exhibited moderate respiratory distress, with the chest radiograms revealing diffuse opacifications consistent with a chemical pneumonitis (Nicogossian 1977; Nicogossian and Campbell 2023), The crew was hospitalized at the Tripler Army Medical Center in Honolulu, Hawaii, remained on steroid and oxygen therapy until the resolution of the pneumonitis. The crew continued their convalescence at the Marine Corps Base Hawaii, located in Kaneohe Bay, for a week before returning to the Lyndon B. Johnson Space Center, Houston, Texas (Nicogossian 1977).

Conclusion

Fifty years after the ASTP, American astronauts and Russian cosmonauts live and work together in low Earth orbit on the International Space Station (ISS). Many international partners from the Canadian, European and Japanese space agencies have joined this partnership. ASTP proved the viability of an androgynous docking system, laying the groundwork for this technology still in use on the ISS program. This system called the *Androgynous Peripheral Attachment System* (APAS) allowed two different spacecraft to dock with one another.

After a symbolic docking of the country's spacecraft, two cosmonauts and three astronauts visited with each other spacecrafts, using a NASA provided transfer airlock, completing five joint experiments and exchanging commemorative items. This mission and collaboration between biomedical personnel and the astronauts and cosmonauts paved the way for many joint research projects culminating in the assembly and operation of the ISS starting 2008. A major contribution of the ASTP to future crewed missions was the development and formulation of joint medical standards and research protocols using a joint U.S./USSR bedrest study as a testbed (Sandler and Grigoriev 1990).

Less publicized were the results of research across different fields scientific, engineering, biomedical, and social disciplines (El-Baz and Warner 1977; Page and Page 1977). Prior to ASTP, NASA robotic missions did include international participation, which was not the case for crewed missions. The exception was the flight of the *Biostack* device, designed to study the biological effects of high energy cosmic radiation [HZE/GCR] were flown on Apollo 16 and 17 missions (Facius et al. 1979, Horneck 1993). ASTP was the first Apollo mission flown at 51.8^{0} inclination to match the orbit of the Soyuz space crafts (Blackmer 1974), which offered an opportunity to follow up on the "light flashes" reported by the Apollo Lunar and Skylab astronauts (Huesman et al. 1976; Tobias et al. 1971) and more recently together with the considered a risk to the vision for interplanetary travel (Aleci 2020).

Summary

ASTP was the first Apollo mission flown at a 51.8° inclination to match the orbit of the Soyuz space craft (Blackmer 1974), which offered an opportunity to follow up on the "light flashes" reported by the Apollo Lunar and Skylab astronauts (Huesman et al. 1976; Tobias et al. 1971) and more recently in combination with the Spaceflight-Associated Neuro-Ocular Syndrome (SANS) considered a risk to the astronaut or cosmonaut's vision for interplanetary travel (Aleci 2020; Brunstetter et al. 2025).

The successful treatment of the chemical pneumonitis was a benchmark for future occupational handling of propellants in the aerospace industry. The increase breakdown of collagen detected after the exposure to N_2O_4 was proposed as a screening test for such exposures (Hatton et al. 1977).

The ASTP was a successful symbol of détente at less than 0.04% of the total Apollo program an estimated cost (Dreier 2020).

Limitations and Bias

Most of the ASTP mission and experiments results were published on government websites. ASTP political success and global benefits beyond cooperation in space has been challenged (Ellis 2019). Compared to other space projects the scientific output is limited since the primary objectives were political and engineering. The majority of the ASTP publications can be found as NASA technical reports and can be found in the NASA STI Repository[1] detailing the engineering, NASA published scientific and policy outcomes of ASTP. While individual findings may have been written up and submitted for peer review in scientific journals, the comprehensive summaries of mission results are found in official NASA reports, which are technical documents rather than traditional peer-reviewed journal articles. The ASTP medical results identify each crewmember by name or crew position and in conformance with the NASA guidelines for release of public information following significant mission anomaly or crew injury.

References

Aleci, C. 2020. From international ophthalmology to space ophthalmology: the threats to vision on the way to Moon and Mars colonization. *International ophthalmology, 40*(3), 775-786.

Brunstetter, T. J., Mason, S. S., Osborne, S., Young, M., Oswald, T., Gibson, C. R., et al. 2025. Predicting Spaceflight-Associated Neuro-Ocular Syndrome in International Space Station Astronauts. *JAMA ophthalmology.*

Devezas, T., de Melo, F. C. L., Gregori, M. L., Salgado, M. C. V., Ribeiro, J. R., and Devezas, C. B. 2012. The struggle for space: Past and future of the space race. *Technological Forecasting and Social Change, 79*(5), 963-985.

Doarn, C. R., Nicogossian, A. E., Grigoriev, A. I., Tverskaya, G., Orlov, O. I., Ilyin, E. A., and Souza, K. A. 2010. A summary of activities of the U.S./Soviet-Russian joint working group on space biology and medicine. *Acta Astronautica*, 67(7-8), 649-658.

Dreier, C. 2022. An improved cost analysis of the Apollo Program. *Space Policy, 60*, 101476.

El-Baz, F., and Warner, D. M. 1977. *Apollo-Soyuz Test Project: Astronomy, earth atmosphere and gravity field, life sciences, and materials processing* (Vol. 1). Scientific and Technical Information Office, National Aeronautics and Space Administration

Ellis, T. 2019. "Howdy partner!" Space brotherhood, détente and the symbolism of the 1975 Apollo–Soyuz Test Project. *Journal of American Studies, 53*(3), 744-769.

1 NASA Technical Reports Server [NTRS]: https://nters.nasa.gov

Facius, R., Bücker, H., Horneck, G., Reitz, G., and Schäfer, M. 1979. Dosimetric and biological results from the Bacillus subtilis Biostack experiment with the Apollo-Soyuz Test Project. In *Life Sciences and Space Research* (pp. 123-128). Pergamon.

Grindeland, R. E., Ilyin, E. A., Holley, D. C., and Skidmore, M. G. 2005. International collaboration on Russian spacecraft and the case for free flyer biosatellites. *Advances in space biology and medicine, 10*, 41-80.

Hall, R. C. 1969. Rescue and return of astronauts on Earth and in outer space. *American Journal of International Law, 63*(2), 197-210.

Hanhimäki, J. M. 2012. *The rise and fall of détente: American foreign policy and the transformation of the Cold War.* Potomac Books, Inc.

Hatton, D. V., Leach, C. S., Nicogossian, A. E., and Ferrante, N. D. 1977. Collagen breakdown and nitrogen dioxide inhalation. *Archives of Environmental Health: An International Journal, 32*(1), 33-36.

Horneck, G. 1993. Responses of Bacillus subtilis spores to space environment: results from experiments in space. *Origins of Life and Evolution of the Biosphere, 23*(1), 37-52.

Huesman, R. H., Upham, F. T., Wieskamp, T. F., Kucala, J. M., Goulding, F. S., Landis, D. A., et al. 1976. 13. Quantitative observations of light flashes sensations experiment MA-106 b. *Apollo-Soyuz Test Project: Preliminary Science Report.*

Israel, M. H. 2012. Cosmic rays: 1912–2012. *Eos, Transactions American Geophysical Union, 93*(39), 373-374.

Kay, W. D. 1998. John F. Kennedy and the two faces of the US space program, 1961–63. *Presidential Studies Quarterly, 28*(3), 573-586.

Krasnyak, O. 2018. The Apollo–Soyuz test project: construction of an ideal type of science diplomacy. *The Hague Journal of Diplomacy*, 13(4), 410-431.

Lewis, J. L., and Donahoe, S. 2023. Space vehicle docking system standardization. *Journal of Space Safety Engineering, 10*(2), 127-132.

Nicogossian, A. E. 1977. *The Apollo-Soyuz Test Project: Medical Report* (No. NASA-SP-411). Pages 7-25, and 48-52 [tremors in lower extremities and fingers].

Nicogossian, A., and Campbell, M. R. 2023. The Legacy of the Apollo-Soyuz Test

Project. *Aerospace medicine and human performance, 94*(8), 651-653.

Page, L. W., and Page, T. 1977. *Apollo-Soyuz [experiments in Space]* (Vol. 133). National Aeronautics and Space Administration.

Ross-Nazzal, J. 2010. Détente on Earth and in space: The Apollo-Soyuz test project. *Organization of American Historians Magazine of History, 24*(3), 29-34.

Sandler, H., and Grigoriev, A. I. 1990. *Joint US/USSR study: Comparison of effects of horizontal and head-down bed rest* (No. A-85177).

Schlesinger, A. 1967. Origins of the cold war. *Foreign Affairs, 46*(1), 22-52.

Tobias, C. A., Budinger, T. F., and Lyman, J. T. 1971. Radiation-induced light flashes observed by human subjects in fast neutron, X-ray and positive pion beams. *Nature, 230*(5296), 596-598.

Weathersby, P. K., Homer, L. D., and Flynn, E. T. 1984. On the likelihood of decompression sickness. *Journal of Applied Physiology, 57*(3), 815-825.

Westad, O. A. 2017. *The Cold War: A world history*. Basic Books.

Zelizer, J. E. 2009. Détente and Domestic Politics. *Diplomatic History*, 33(4), 653-670.

This publication is available open access at:
http://www.ipsonet.org/publications/open-access

Thanks to the generosity of the American Public University System

APU American Public University

AMU American Military University

Related Titles from Westphalia Press

The Limits of Moderation: Jimmy Carter and the Ironies of American Liberalism

The Limits of Moderation: Jimmy Carter and the Ironies of American Liberalism is not a finished product. And yet, even in this unfinished stage, this book is a close and careful history of a short yet transformative period in American political history, when big changes were afoot.

The Zelensky Method by Grant Farred

Locating Russian's war within a global context, The Zelensky Method is unsparing in its critique of those nations, who have refused to condemn Russia's invasion and are doing everything they can to prevent economic sanctions from being imposed on the Kremlin.

Sinking into the Honey Trap: The Case of the Israeli-Palestinian Conflict by Daniel Bar-Tal, Barbara Doron, Translator

Sinking into the Honey Trap by Daniel Bar-Tal discusses how politics led Israel to advancing the occupation, and of the deterioration of democracy and morality that accelerates the growth of an authoritarian regime with nationalism and religiosity.

Essay on The Mysteries and the True Object of The Brotherhood of Freemasons by Jason Williams

The third edition of Essai sur les mystères discusses Freemasonry's role as a society of symbolic philosophers who cultivate their minds, practice virtues, and engage in charity, and underscores the importance of brotherhood, morality, and goodwill.

Bunker Diplomacy: An Arab-American in the U.S. Foreign Service by Nabeel Khoury

After twenty-five years in the Foreign Service, Dr. Nabeel A. Khoury retired from the U.S. Department of State in 2013 with the rank of Minister Counselor. In his last overseas posting, Khoury served as deputy chief of mission at the U.S. embassy in Yemen (2004-2007).

Managing Challenges for the Flint Water Crisis Edited by Toyna E. Thornton, Andrew D. Williams, Katherine M. Simon, Jennifer F. Sklarew

This edited volume examines several public management and intergovernmental failures, with particular attention on social, political, and financial impacts. Understanding disaster meaning, even causality, is essential to the problem-solving process.

User-Centric Design by Dr. Diane Stottlemyer

User-centric strategy can improve by using tools to manage performance using specific techniques. User-centric design is based on and centered around the users. They are an essential part of the design process and should have a say in what they want and need from the application based on behavior and performance.

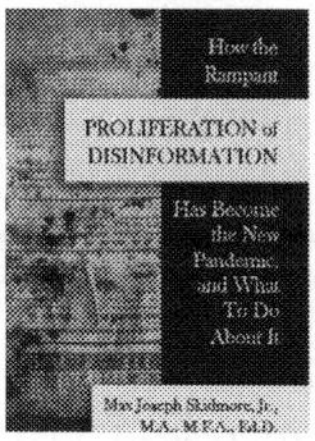

How the Rampant Proliferation of Disinformation Has Become the New Pandemic, and What To Do About It by Max Joseph Skidmore Jr.

This work examines the causes of the overwhelming tidal wave of fake news, misinformation, disinformation, and propaganda, and the increase in information illiteracy and mistrust in higher education and traditional, vetted news outlets that make fact-checking a priority.

Abortion and Informed Common Sense by Max J. Skidmore

The controversy over a woman's "right to choose," as opposed to the numerous "rights" that abortion opponents decide should be assumed to exist for "unborn children," has always struck me as incomplete. Two missing elements of the argument seems obvious, yet they remain almost completely overlooked.

Made in United States
Orlando, FL
04 November 2026

90218652R00046